The Ph.D. Code

Tomasz Liskiewicz · Grzegorz Liskiewicz

The Ph.D. Code

60 Tips to Get your Degree in STEM Subjects

 Springer

Tomasz Liskiewicz
Department of Engineering
Manchester Metropolitan University
Manchester, UK

Grzegorz Liskiewicz
Faculty of Mechanical Engineering
Łódź University of Technology
Łódź, Poland

ISBN 978-3-031-90101-0 ISBN 978-3-031-90102-7 (eBook)
https://doi.org/10.1007/978-3-031-90102-7

© The Editor(s) (if applicable) and The Author(s), under exclusive license to Springer Nature Switzerland AG 2025

This work is subject to copyright. All rights are solely and exclusively licensed by the Publisher, whether the whole or part of the material is concerned, specifically the rights of translation, reprinting, reuse of illustrations, recitation, broadcasting, reproduction on microfilms or in any other physical way, and transmission or information storage and retrieval, electronic adaptation, computer software, or by similar or dissimilar methodology now known or hereafter developed.
The use of general descriptive names, registered names, trademarks, service marks, etc. in this publication does not imply, even in the absence of a specific statement, that such names are exempt from the relevant protective laws and regulations and therefore free for general use.
The publisher, the authors and the editors are safe to assume that the advice and information in this book are believed to be true and accurate at the date of publication. Neither the publisher nor the authors or the editors give a warranty, expressed or implied, with respect to the material contained herein or for any errors or omissions that may have been made. The publisher remains neutral with regard to jurisdictional claims in published maps and institutional affiliations.

Cover image by Freepik

This Springer imprint is published by the registered company Springer Nature Switzerland AG
The registered company address is: Gewerbestrasse 11, 6330 Cham, Switzerland

If disposing of this product, please recycle the paper.

Contents

1 **Own the Ph.D. Process** 1
   1.1 Tell Me Why Are You Doing This? 1
   1.2 Planning for a Success 3
   1.3 You and Your Supervisor 7
   1.4 You Are Part of the Team 9
   1.5 Why Bother with Meetings? 11
   1.6 The End of Your First Year 13
   1.7 Battling the Inner Critic 15
   1.8 Writing Thesis 17
   1.9 Preparing for the Viva 19
   1.10 Is There Life After Ph.D.? 22

2 **Understand and Organise the Context** 25
   2.1 Strategies for Managing Your References 25
   2.2 Staying on Top of the Latest Literature (in a Smart Way) ... 27
   2.3 Can State-of-the-Art Be Two-Dimensional? 29
   2.4 Tips for Organising Your Data 32
   2.5 Natural Planning Method 35
   2.6 Being Creative in Your Research 37
   2.7 Investing in Yourself 40
   2.8 Racing with Technology 42

3 **Craft Impactful Research Papers** 45
   3.1 Why Do You Want to Publish? 45
   3.2 What's Your Story? 47
   3.3 Your Title Can Make or Break Your Paper 50
   3.4 How to Write a Captivating Abstract in Ten Minutes 51
   3.5 Writing Your Introduction in Seven Easy Steps 53
   3.6 Let's Have a Meaningful Discussion 55
   3.7 Nine Ways to Break Your White Page Syndrome 57
   3.8 Plagiarism in Scholarly Publishing 59
   3.9 Are You a LaTeX Geek? 61

4 Beyond the Draft: Get Published ... 63
- 4.1 Understanding Peer-Review ... 63
- 4.2 The Best Journals for Your Research ... 65
- 4.3 Who Are Your Co-authors? ... 67
- 4.4 High-Quality Diagrams and Figures ... 70
- 4.5 Cover Letter ... 72
- 4.6 Suggested Reviewers ... 75
- 4.7 Submitting Your Paper ... 76
- 4.8 Reply to Reviewers Only Once ... 79

5 Master Conferences and Presentations ... 83
- 5.1 Preparing for the Conference ... 83
- 5.2 Making Meaningful Connections at the Conference ... 85
- 5.3 Five Steps to an Impactful Presentation ... 88
- 5.4 Three Constraints That Shape Your Presentation ... 90
- 5.5 Getting the Audience Crazy About Your Content ... 92
- 5.6 Why Body Language Really Matters ... 95
- 5.7 When to Stop Practising and Start Delivering ... 97

6 Become Credible ... 99
- 6.1 Why Do I Need to Promote Myself? I'm a Researcher, Not an Influencer ... 99
- 6.2 Tools to Make Your Research More Visible ... 101
- 6.3 Academic Social Networks—ResearchGate and Academia.edu ... 104
- 6.4 Professional Social Networks—LinkedIn and X ... 105
- 6.5 Your Profile Across Platforms—from ORCID to Personal Page ... 108
- 6.6 Research Paper Repositories ... 110
- 6.7 Popular Science Communication ... 114
- 6.8 Stepping into the Reviewer's Shoes ... 116
- 6.9 Should I Consider Research Mobility? ... 119

7 Develop Personal Toolkit for Success ... 123
- 7.1 Your Morning Routine Is Your Friend ... 123
- 7.2 The Only High-Performance Supplement You Need ... 124
- 7.3 Habits, Habits, Habits ... 126
- 7.4 Small Wins, Big Impact ... 128
- 7.5 Mindfulness ... 129
- 7.6 Eat the Frog First ... 131
- 7.7 The Pomodoro Technique ... 133
- 7.8 Do You Work or Do You Create? ... 134
- 7.9 Choosing What Truly Matters ... 136

About the Authors

Tomasz Liskiewicz is Professor of Tribology and Surface Engineering at Manchester Metropolitan University, where he also serves as Head of the Department of Engineering. With over 25 years of international academic and engineering experience, Tomasz has held positions at leading research institutions in the UK, France, Canada, and Poland. His research focuses on the tribology and surface engineering of functional surfaces, with particular attention to engineering coatings, their architectures, and their optimisation for enhanced performance.

Tomasz holds a Ph.D. in tribology from École Centrale de Lyon in France and an M.Sc. in materials science from Łódź University of Technology in Poland. He has extensive practical experience, having spent two years as Senior Scientist at Charter Coating in Alberta, Canada, where he led material testing projects for the oil and gas industry.

In addition to his research, Tomasz has taught undergraduate, M.Sc., and Ph.D.-level courses in tribology, surface engineering, and fluid dynamics. He has a track record of supervising Ph.D. students, providing them with expert guidance in their academic and research activities, supporting their development into independent researchers. Together with Grzegorz, he developed and delivered a series of workshops aimed at early-career researchers.

Tomasz was elected as Fellow of the Institution of Mechanical Engineers in 2014 and Fellow of the Institute of Physics in 2019. He is Editor-in-Chief of *Tribology—Materials, Surfaces and Interfaces*, a journal published by Sage, and runs a successful not-for-profit initiative, Surface Ventures, which provides surface engineering webinars for both academics and industry professionals. Additionally, Tomasz serves as Chairman of STEM Racing Poland. He loves discovering the intricacies of mechanical watches and is always planning his next travel adventure.

Grzegorz Liskiewicz is University Professor at Łódź University of Technology. His research interests are focused on turbomachinery, industry 4.0, energy, design and safety of centrifugal compressors. He worked at the Texas A&M, Leeds University, Oxford University, Cambridge University, University of Strathclyde, and General Electric. Greg graduated his Ph.D. jointly from the University of Strathclyde and

Łódź University of Technology with an honorary degree. Greg is Alumni of the Top 500 Innovators Poland Program organized by Polish Ministry of Science, the International Visitor Leadership Program organized by the U.S. Department of State and the Leadership Academy for Poland organized by the Centre for Leadership. He worked as Board Member of the university tech transfer office, the Rector's proxy for academic entrepreneurship, and served as Mentor for startups and young researchers. In addition to his research, Greg has taught undergraduate, M.Sc., and Ph.D.-level courses in programming, compressor design, and writing academic papers. He has a track record of supervising Ph.D. students in Poland and UK, supporting their growth both: professionally and personally to become independent researchers or start successful industrial careers. He delivered a series of workshops aimed at early-career researchers in Poland, UK, Estonia, and USA. Grzegorz serves as Vice-Chairman of STEM Racing Poland. In a free time, he reads books and listens to vinyl discs. His favourite way of travelling is by bike, and the favourite sport is ice hockey.

Getting Started

Is This Book for Me?

If you're reading this, it means you're probably part of a wonderful group of people who have chosen to dedicate a great part of their life to the pursuit of knowledge and bettering the world. So, before we start, let's first appreciate where you are and how much you can take from the Ph.D. experience. Strap in, it's going to be one hell of a ride.

There'll be times where you struggle, times where you'll hit a seemingly unconquerable wall. But equally, there'll be times where you make this little difference in the world and remind yourself why you choose this path in the first place.

So, is this book really for you? Our assumption is that you are somewhere in the beginning of your Ph.D. adventure. But we believe that the book can help anyone at the early stage of scientific research and academia: from undergraduate to post-doc. All the lessons we share come from our own lived experiences. We both were once just like you.

We are engineers, and this book was written with Ph.D. students in STEM fields in mind. However, most of the topics discussed apply to Ph.D. students across all scientific disciplines.

Academia is full of unwritten rules and secret rituals. We are like a tribe in some way, and we don't only mean that the university people love strange ceremonies where everyone wears full academic dress (but it surely looks odd for an outsider). We also mean a number of terms & conditions which are not even written in small print on your contract. You just learn them with time, that's the consensus—obviously also unwritten. You master collaboration with the team and supervisor, set up your research activities, publish papers, go to conferences and workshops, work on your outreach and productivity routines. Finally, after writing your thesis and defending your Ph.D., you are admitted to one of those secret rituals, where someone will give you the diploma (and depending on the country, you may be symbolically struck with an academic hat or sceptre).

What if someone had told you, right from the start, how things really work? It could have saved you countless hours of frustration and made the entire journey far more enjoyable. That's exactly how we believe research should be—hard work infused with excitement and a sense of discovery. That belief is what inspired us to create this book. It's the resource we wish someone had handed us when we stood at the starting line of our Ph.D. journey. Nobody did, but we were fortunate to have incredible mentors who guided us along the way. Now, it's our turn to give back. This book is your guide to navigating the highs and lows of the Ph.D. experience. It's an invitation to join the academic tribe—no admission fee, no hidden rules, and no dark secrets.

How Was This Book Written?

This book features 60 concise sections, each focusing on a key aspect of the Ph.D. journey. Every section is rooted in real recommendations. We reached out to former Ph.D. students, colleagues, and coworkers, asking them to share the pain points they encountered during the process. Their insights shaped the responses we gathered, forming the foundation for this book's table of contents.

Each section of this book was written by one of us, then reviewed by the other, before being sent back for revisions. We repeated this process several times, going back and forth until we were satisfied with the result. We've lost count of the iterations, but every cycle was aimed at ensuring we both fully agreed on the final version. In this way, each section reflects the combined experiences of both of us. However, we chose to use the phrases "I" and "me" throughout the book, as we believe it makes the narrative more engaging and relatable. In this context, "I" represents a collective voice, blending our diverse perspectives, shaped by the cultures we've encountered, our backgrounds, our personalities, as well as our shared passion for academia and engineering. This "collective I" approach will take effect starting from the next paragraph!

How to Make the Most of This Book?

Okay, before I actually dive in, I want to give you some pointers on how to make the most of this book. Over the course of seven chapters, there are no less than 60 sections filled to the brim with tips. This book is written in simple terms, breaking down some of the most complicated and sometimes bizarre sides of being an early-stage researcher into digestible pieces of advice you can apply to your work today.

And that's a lot to take in. So, I suggest two possible strategies.

- **Whole-at-once**—if you want to give yourself an oversight, you can go through the whole content. It doesn't necessarily need to be in order, as each section is quite independent.
- **At-demand**—the book could be used as and when you need it. I want it to be the handbook you turn to when you need some help or some motivation to get out there and get things done.

Each Chapter talks about a different aspect of your academic life:

- **Chapter 1**—An overview of your Ph.D. from an organizational and formal point of view going in chronological order—from setting up your work and relationship with supervisor, through all kinds of meetings, reviews and ending up with the viva and your life after Ph.D.
- **Chapter 2**—Let's now turn to the substantive core of your Ph.D. This chapter will be a mix of reflections and strategies on managing knowledge and using it for your Ph.D. work. It includes pieces like being on top of the state-of-the-art, organising papers and staying creative.
- **Chapter 3**—Writing academic papers—that's something you cannot omit in your Ph.D. I discuss the key sections of a paper and help you in getting through the process.
- **Chapter 4**—Writing paper is just a half-marathon. To make it a full distance, you need to get the paper published. For that you need to understand how journals work and how to make them work in your favour.
- **Chapter 5**—All about conferences and another external activities you can undertake. How to prepare for them. How to prepare a great presentation and make meaningful connections during the events you participate in.
- **Chapter 6**—If writing and publishing is a full-marathon, then making your paper known and recognised by community is like triathlon. In this chapter I discuss how to outreach your research activities efficiently.
- **Chapter 7**—Intellectual work can be overwhelming. To cope with this, I am suggesting techniques and advice on how to be productive and treat yourself right.

Want to know how to write each section of your dissertation? Head on over to Chap. 4. Going to your first conference? Then Chap. 6 is for you. As you progress, you might be hopping back and forth between chapters to find advice that's relevant to you in the present moment.

This book will help you become a better writer, communicator, researcher, and colleague. Your Ph.D. journey isn't just about conducting research and writing to the best of your ability. It's also a unique opportunity to build meaningful connections and lay the groundwork for a thriving career and a fulfilling life.

Most importantly, I don't want you to feel as though this book is judging you. It might sound corny, but I want it to feel like a supportive friend. At times, it will encourage you to be introspective and reflect on who you are, but this isn't criticism—it's an invitation to grow and improve. Personal growth is a lifelong journey, one we must all undertake. And while the path is never-ending, it's one we can walk together and find joy in along the way.

Sometimes, I'll provide you with a straightforward to-do list, almost as if no objections are allowed. The reason is simple—lists are the most concise way to deliver information. As an engineer, I aim to optimise the density of useful information per number of words. But let me be clear: these aren't orders. Please treat them as advice, shared from experience. This book is a starting point, a guide to help you find your way of doing things. But it's your journey, and your way will always be the best way!

Enjoy it!

Chapter 1
Own the Ph.D. Process

> The moment you stop driving the process is the moment the process starts driving you.

1.1 Tell Me Why Are You Doing This?

A Little Bit of Philosophy (Coming from an Engineer)

Working on a Ph.D. is truly one of the most transformative challenges you can undertake at the start of your career. If you're reading this book, you probably think so too. But let me share my (somewhat philosophical) reflection on it.

When you work in a company, someone pays you to devote your time to do the work. I see it as selling time, experience, and skills to someone in exchange for a salary. The employer buys them from you in order to realise their own vision and goals. You're rewarded handsomely, but in return, someone decides what you should do. This arrangement is very suitable in many circumstances. If the vision of the company is consistent with yours, you have the opportunity to join an established organisation and take advantage of the opportunities it offers.

If you don't like the idea of "selling your time", you can use it for your benefit. People who start their own company or start-up, invest their time, experience, and skills to achieve their own goals. Such investment, like any other, brings high risk. It can turn out to be a bullseye and bring staggering results in the future. But the investment may also fail, and you won't achieve the desired results. In any case, this approach gives you an opportunity to create something of your own and on your terms. You create your own vision according to your beliefs.

Viewing your time and work in this way raises an interesting question: how does a Ph.D. fit into this framework? A Ph.D. is a unique experience that provides a high degree of independence. Essentially, you are being paid to dedicate yourself to solving a single problem. At the same time, you are also having the opportunity to develop personally and professionally, increasing the value of your work and enhancing your future career prospects. Sounds amazing, right? However, this freedom doesn't mean you have complete autonomy during your Ph.D. You will likely have well-defined objectives tied to your thesis and may also take on additional responsibilities, such as teaching or fulfilling obligations required by the sponsor of your research project.

There will also be rules—sometimes unwritten—that impose additional tasks. As a result, you may encounter moments when you're not entirely free. However, you're in a truly fortunate position—one that offers a unique sense of intellectual freedom and creativity. This wide independence, combined with the opportunity to focus on your own vision and personal development, is a comfort many people can only dream of!

Lesson from One of the Best—Albert Einstein

To prevent your freedom from working against you, it's crucial to clearly visualise how you will approach the task from the very beginning. Set clear goals and establish a structure that will guide your work throughout the process.

Let's start with a scientific example—from the work of Albert Einstein himself. He once created a formula to describe the movement of a large particle when surrounded by many smaller molecules—the so-called Brownian Motion. Scottish biologist Robert Brown observed the movement of pollen suspended in water. He noted that they move in a chaotic way. Albert Einstein suggested the reason for this could be the fact that the pollen is constantly being bombarded by smaller particles. Ones that weren't visible through the microscope (at then available magnification level)—water particles.

He derived a mathematical model which shows that a randomly moving object moves away from its starting position proportionally to $\sqrt{t}$. If after 1 min the particle reaches a distance of 1 cm, after 4 min it reaches 2 cm, after 9 it gets to 3 cm. After 100 min, the particle reaches 10 cm.

Now, for comparison, take a particle that moves in a straight motion in one specific direction, but with 5 times less speed. After one minute, it will reach a distance of 0.2 cm. After 9 min the distance will be 1.8 cm. After 100 min, however, it will be at a distance of 20 cm. And, of course, its advantage will only increase further with time.

The above example shows a simple difference between chaotic wandering and persistent movement in a set direction. The latter example is more effective in the long run, even if you're moving at a much slower speed. The same is true for your progress over the coming years. If you are subject to chaotic impacts coming at you from all angles, you will move much slower (even if you work crazy hard to maximise the daily progress). So instead, you want to be a Ph.D. student that moves in a straight, conscious manner. I will go back to the importance of keeping the constant direction close to the end of this book—in Sect. 7.4.

Many people blame their insufficient efficiency on not working hard enough. They believe that in order to achieve the desired results, they have to work more effectively and keep working more hours per day. This can obviously be helpful, but in the long run, the effects can be devastating. How many days can you actually work *well* at full speed? In fact I can argue that doing less can lead to greater results in a long run, as highlighted by Cal Newport in his book "Slow Productivity".

The conclusion is straightforward: to achieve long-term productivity, it is essential to clearly understand the purpose of your Ph.D.

Know the "Why"

The key question then becomes: "Why are you doing this?"

Don't mix it with another important question: "What are you going to do?" which is about defining the scope of actions you're going to undertake. For example, the sentence "I am going to investigate how the physical activity of the patients affects their chances of success in treatment of disease X" is a range of actions. It doesn't answer why such research is needed. Or why it will be valuable.

The key "why" question is also not about determining how you are going to achieve your goal. Someone could say "I am going to do a set of surveys combined with monitoring the progress of treatment of disease X". The choice of how you will conduct your research: the method, apparatus, reagents, and possible statistical analyses is very important, but it still not the question that should be asked first.

The good "why" questions are: "Why do you want to tackle this specific issue or piece of research?", "Why is it important?", "Why is it valuable to yourself or others?" The answer may be "I want more people to be successfully treated for disease X". The question "why" touches much deeper level of awareness and understanding. It concerns the core motivation for your research. The importance of asking the right question was described by Simon Sinek in a book "Start with Why".

This doesn't mean that the research must be commercially applicable. The "why" is not necessarily related to the socio-economic arguments. The "why" may be scientific in nature and, therefore, be related to the desire for a deeper understanding of a phenomenon or topic. But it's great to know the "why" of your Ph.D. By knowing it:

- You will better organise your work aiming at the expected result.
- You will know which results are crucial for the final outcome.
- Your publications will much better present the impact of your work.
- You will have more enthusiasm and motivation for your work.

Takeaway points:

- **Ph.D. is a unique job: it gives you intellectual freedom.**
- **Moving steadily in one direction is better than rushing aimlessly.**
- **The best way to understand purpose of your Ph.D. is to answer the "why" question.**

1.2 Planning for a Success

Set the Milestones

Early on, your supervisor will likely introduce you to your research topic. It's crucial to understand your supervisor's expectations—where they envision you at various stages: one year in, midway through, and at the finish line. What are the key milestones for the success of your thesis? Once you have a clear understanding of these expectations, you can create a focused action plan to guide your progress.

The plan should include so-called "milestones"—tangible and verifiable results that should be achieved at a given stage. Milestones are a powerful tool in planning a research project. If you define the planned course of the project at the beginning, the milestones will allow you to assess whether your pace of work is right and if you're on track to finish your work on time.

When humanity was based on physical work, everything was easily measurable. The farmer had to do the work on a certain area of the field. He knew perfectly well when he reached the goal. A researcher doesn't have this luxury—Intellectual work cannot be measured that easy. There's always something that could be done better. There's always another hypothesis that you could investigate. There's always the possibility of taking additional measurements. Your Ph.D. thesis can always have one more chart, equation, or chapter.

However, the duration of your Ph.D. scholarship, is clearly defined. You must realistically adapt your plan to this timeline. It's essential to plan what results you hope to achieve at key milestones throughout your Ph.D. to ensure you stay on track and make steady progress.

A good milestone should meet the SMART principle:

- **Specific**—simple to verify and significant for achieving the final result.
- **Measurable**—you can say whether it's achieved or not by measuring some quantity.
- **Achievable**—something you can actually achieve in such a timeframe.
- **Relevant**—consistent with the objective of the final Ph.D. thesis
- **Time bound**—with clearly defined completion point in time, taking into account possible delays.

Here are some specific examples of good Ph.D. milestones:

- Within the first nine months, complete at least 10 measurement series (that include predefined quantities you've listed).
- By the end of the first year, finalise technical drawings for the test stand in a format suitable for manufacturing by an external company.
- By the end of month 15, present a report showing numerical simulation results that demonstrate at least 10% compliance with experimental data (with clearly defined parameters for compliance).
- Submit a research paper on topic X to journal Y by deadline Z. The preparation of a paper inherently involves a well-structured set of tasks and significant intellectual input, making it an excellent milestone.
- By the end of month 30, finalise all lab experiments and shift focus from laboratory work to writing and analysis.

Examples of less effective milestones:

- **A complete literature review**—A literature review is never truly "complete." There are always additional relevant papers waiting to be discovered. Author needs to specify what criteria the literature review must meet, to be considered useful for the next stages of the work. A better milestone would be: "I will create a literature review matrix with at least X high-quality entries" (see Sect. 2.3 for more on the literature review matrix). While this still isn't perfectly specific—since the number of references doesn't guarantee their usefulness—it encourages you to think critically about the quality of each entry you include.
- **Report on the results of the research**—Let's be honest, this type of milestone can be easily met with any form of written text. Could a five-page report written overnight qualify as a "report on results of the research"? Absolutely. Instead, consider what specific features the report should have to make it truly useful. Focus on measurable aspects that can be validated at this stage. While the number of pages might be a factor, remember that this isn't necessarily a "the more, the better" situation. Consider outlining the key features and content required for the report at this stage.

Before we move on, I have one final comment on measurability. It's important to have at least one measurable component for each milestone. However, remember that *"not everything that can be counted counts, and not everything that counts can be counted"* (a quote often attributed to Albert Einstein). Therefore, feel free to incorporate elements that highlight other crucial, though not easily measurable, aspects of your work. A good example is the inclusion of "good quality entries" in the literature review milestone. While quality is difficult to measure, it is fundamentally important to the success of your work.

Mike Tyson Gives a Good Lesson for Project Planning

You already have your plan. Now all that's left is to implement it, point by point, to achieve it perfectly, right?

No, unfortunately it won't be that easy. It has never happened to me that a research project spanned over several years has gone completely according to plan. I therefore suspect that in your case, too, this may not work.

Let's turn to wise words of a champion boxer Mike Tyson: *"Everyone has a plan until they get punched in the face"*. Fortunately, research work is not a fight, but it can certainly bring about unexpected "punches". The equipment may break down, the results may not match the theory, the delivery of samples may be delayed—and other such haunting examples. So, get ready for unexpected events and the changes they cause. You need to be flexible and regularly adapt your plan to reality.

Unexpected events are part of a scientific work. Albert Einstein said, *"If we knew what we were doing, this wouldn't be called research"*. Whenever change happens, think about the milestones. These are key elements you have defined initially as crucial for delivering the final result. Will you still be able to reach them on time? If not, then just think of the plan B and re-iterate the plan.

A Gantt Chart? Really?

Yes, really! A Gantt chart might seem like a tool better suited for construction sites or corporate project management, but it can be incredibly valuable in academic research as well. In essence, a Gantt chart is a timeline that visually represents tasks, their duration, and their dependencies. This means that you can map out your tasks, milestones, and deadlines in a way that gives you a clear overview of your progress and workload.

A Gantt chart allows you to visualise the entire project at a glance. Ideally on one page! With it, you can immediately identify overlapping activities or periods of intense workload. Many research activities are dependent on others that need to be completed first. For example, you might need to complete data collection before moving to the analysis. A Gantt chart makes these dependencies clear.

As I said, academic research often involves tasks that are unpredictable or difficult to estimate in duration. Sometimes you get punched! So don't treat your Gantt chart as rigid structure. Too much focus on perfect adherence to the chart can lead to frustration when inevitable changes occur. Once they indeed happen, Gantt chart helps you adjust your timeline. You can quickly see how shifts in one task affect others and recalibrate your project accordingly.

Creating a Gantt chart doesn't have to be complicated. You can use simple tools like Microsoft Excel, Google Sheets, or specialised software such as Trello, Asana, Canva or dedicated Gantt chart programs like Microsoft Project or GanttProject.

Here are some ideas how to set a Gantt chart up for your Ph.D.:

- **List your tasks** —Start by identifying all the major tasks in your research. Break them down into smaller, manageable components, such as "Conduct literature review," "Design experiment," "Collect data," "Analyse results," and "Write thesis chapters."
- **Define timelines**—Assign start and end dates to each task, keeping in mind the dependencies between them.
- **Allocate resources**—If you need specific resources—like lab access, equipment, or supervision—ensure these are planned into your timeline.
- **Add milestones**—Mark your key milestones on the chart for easy reference. Typically, milestones are located at the end of a given task (to make sure it gave satisfactory results), or at the beginning of a task (to make sure you have all resources and data to start it). Locating a milestone in the middle of a task can also be a good way to track its progress.
- **Regularly update**—Use the chart as a living document, updating it whenever changes occur in your plan.

And here are some common mistakes students make when preparing their Gantt charts.

- Forget about public holiday and breaks.
- Lack of a project risk assessment.
- Contingency plan in case your supervisor is unavailable for certain period, e.g. three months.

- No dedicated time to write research papers.
- Not sufficient time for writing up and revising your thesis.

A Gantt chart may not eliminate all the chaos of research, but it provides structure and clarity. It's a visual representation of your path to success and a reminder of what needs to be done. Remember, like any tool, its effectiveness depends on how you use it. So, start simple, stay flexible, and let the chart evolve with your Ph.D. journey.

Takeaway points:

- **What gets measured gets managed. What cannot be measured can be also important.**
- **Make a plan and set milestones that follow the SMART rule.**
- **Visualise your plan using a Gantt chart. Treat it as a living organism and let it evolve with your Ph.D.**

1.3 You and Your Supervisor

Friendship Is Not Obligatory, Professional Communication Is

You should prioritise establishing a clear line of communication with your supervisor early on. Over the next few years, you'll work as part of their team and carry out research in their lab. It's important for you to understand what their expectations of you are. Do you know what question your research work should answer, the already-mentioned "why"? The answer to this question determines your future—make sure that both: you and your supervisor define this answer in the same way.

I'm not saying your supervisor will be your friend (although this sometimes happens). In the beginning, you should focus on building a professional relationship based on understanding and effective communication. Their goal is to carry out research at the highest possible level. If that's also your goal, then you have something in common. You can form a professional relationship and communicate effectively.

Sometimes things go south in the relationship with your supervisor. This can lead to a breakdown in communication and pose your long-time collaboration at risk. Managing such a situation requires a combination of professionalism and resilience. It is best to start by identifying the root causes of the conflict, whether they stem from misaligned expectations, differing communication styles, or a lack of mutual understanding. My suggestion is to focus on communication which often causes most trouble. You can seek advice from trusted colleagues, mentors, or the university's support services to gain perspective and make space for more reflection. In extreme cases, exploring a change of supervisor through appropriate institutional channels may be necessary, but this step should be taken thoughtfully and as a last resort. Before going that far, let's look into potential challenges connected with communication.

Communication Is Key

Setting regular meetings is the most convenient way to keep communication alive. It's very important to keep your supervisor updated about the progress of your work. Take this opportunity to show that you are committed to your research and there is a steady progress. Apart from regular meetings, it can be achieved by sending brief email updates. And remember about the other side—respond quickly to supervisor emails, even if the answer is just "OK" or "thank you". The absence of an answer leaves them with a doubt whether you are doing well.

Be proactive in your communication. If you find yourself in need of help, make sure to inform your supervisor. If you have an interesting idea, don't hesitate to share it. A supervisor doesn't always have the time or may not always take the initiative, sometimes intentionally. Many supervisors believe that their Ph.D. students should be independent from day one. However, in any case, you are fully entitled—and I would even say, obliged—to request a meeting whenever you need support.

One common issue with meetings is the lack of agreement on the conclusions. At the end of a meeting, you may have a clear understanding of the key outcomes, but the question is: does your supervisor share the same conclusions? If not, a potential disaster is brewing. To avoid this, document all key decisions and action points, and summarise them at the end of the meeting. You can also send a follow-up email to confirm these points. To do this effectively, you'll need good meeting notes, a subject covered in Sect. 1.5.

Also, remember to maintain a positive attitude. Your supervisor may ask for your help, whether it's to teach while they attend a conference, run a student project, or find relevant papers for the research. While these requests may not always align with your plans—and you might already have a lot on your plate—keep in mind that helping in such situations will build trust between you and your supervisor. It's an opportunity to demonstrate your professionalism and enthusiasm. When your supervisor sees this side of you, it could lead to more exciting and rewarding tasks in the future.

Supervisor Versus Mentor

If your relationship develops, your supervisor may become your mentor. In order not to confuse these terms, let's describe briefly how we understand them.

Supervision is a duty. It is most often coming from the organisational structure. The role of a supervisor is to work with the individual, and ensure they have the necessary resources and perform their role correctly.

Mentoring is a relationship. In this case—between an experienced professional and a person at the beginning of their career. It is based on mutual trust and respect, a mentor shares their experience, and the mentee shares their energy, creativity, and enthusiasm. Mentoring may be in line with the structure of the organisation, but it is not required. For a mentoring relationship to develop between the mentor and the mentee, the willingness of both parties is needed.

Your mentor may be another academic who is not your supervisor, or even an experienced person from outside of your institution. You may also have more than one mentor. Someone may be your mentor in the field of research, someone in the field of public speaking, and someone in the field of education. You can never have too many mentors!

How does someone become your mentor? Most often, it happens naturally; there's a mutual understanding and respect, which is the foundation for building this relationship. You can also go out and find one yourself. There are so-called 'mentorship programs' where mentors and mentees are paired by the organisers. Just look around—it's likely that your university, or another organisation of your interest is running such a program.

Another way is to just ask. If you would like to be mentored by someone you truly admire, there's nothing to prevent you from asking for support. I would follow a Tim Ferriss advice on how to ask for a mentorship in a really good way:

- **Approach this person individually**—Even if it's just an e-mail, you can make it personal by writing why you decided to ask that specific individual.
- **Remember that mentorship is a relationship**—It works both ways. Explain what you can offer in return for their support.
- **Describe your expectations**— You should explain what kind of support are you looking for, and what time commitment are you expecting.
- **Leave all options**— Be mindful to create a peaceful and respectful space in which "no" can be an acceptable response. Even if the response will be negative, it might still bring some positives: they can make an interesting introduction or you can stay in touch for possible future collaboration.

Takeaway points:

- **Make sure from the very beginning that you agree with the supervisor on your Ph.D. goals.**
- **You are co-responsible for effective communication with your supervisor**.
- **Your supervisor might become your mentor. It's also fine to have additional mentors**.

1.4 You Are Part of the Team

A Ph.D. Is a Team Sport

I sent the initial plan of this book to fellow Ph.D. students and asked them for their opinions. One of them surprised me very much: "Be sure to tell the reader that a Ph.D. is a team thing!"

Indeed, it seemed obvious to me, but it's worth devoting a whole section to this topic. Researchers are smart people, looking for their own answers—they're often individualists. And yet, success in research is still the result of a teamwork.

We work in increasingly large teams. The number of authors of scientific publications is growing regularly, and in some disciplines, it can be counted in thousands (search: Nature, paper authorship goes hyper). Of course, there are some disciplines in which single authorship is still a common thing. But even in these instances, collaboration with others may be necessary (and will certainly bring you many benefits).

A Ph.D. thesis is a perfect example of such case. It is, of course, a single author's thesis and you're responsible for the final result. However, the path to this result will involve a great deal of teamwork: cooperation in a research group, group discussions, joint projects and papers.

Understand the Group Structure

You're a member of a research team. To start with, get to grips with its formal and informal structure. The formal structure in universities can be very confusing. At the beginning, try to get an idea of the roles assigned to people within the research group. This often overlaps with the roles in the faculties, the school, and the university. You can learn about the structure at each of these levels on the website. It will give you an idea of who you can contact for formal matters such as:

- Accessing the computing server.
- Using the laboratory equipment.
- Acquiring signatures for travel approvals.

The informal structure is much more difficult to figure out. It results from institutional habits and relations between people. An important function in an informal structure is someone I call here "champion". This position is given (in an non-official way) to someone who has proven perfection in a certain area. It can be a champion in a specific research discipline, but also in skills that are important elements of a researcher's toolkit (programming, statistical analysis, writing scientific publications, mathematical modelling, scientific presentations), or other skills (visionary, networker, creative guru).

Some universities recognise the importance of such a person and also formalised this function. Another good name would be the "gatekeeper"—someone who keeps the key to the most important problems you might face on the way.

It's worth figuring out everyone's informal roles—who they are in the group. Don't be afraid to ask your colleagues who's best placed to address a given problem. When you knock on this person's door, just say: "I heard you're the expert on X, I need help—could we meet for a coffee?"

If you use someone else's help, remember to thank them for it. If someone's help makes an impact—be it in a presentation, poster, or paper—consider putting the person's name in the acknowledgments section. If it isn't that major, it's still worth sending them a thank you email or going out of your way to shake their hand for their assistance.

Find Your Spot

So now you know the formal and informal structure in your group. You know the champions in key areas and the people who provide particular value to the team. The question now is where do you fit into that group? If you want to be an important part of the team, it's worthwhile trying to develop a role in the informal group structure. If you want to do this, you need to think about what you want to be known for.

The answer to this question depends on two factors. The first is your passion. If there's a topic that particularly fascinates you, the role of "champion" will be very fulfilling. Imagine, without any effort on your part, people come to you with questions or ideas about something that's particularly interesting to you. Sounds good, right?

The second factor is where are the needs of the group itself. Perhaps there are areas that are particularly needed by the team. There's always some kind of research area gaining popularity. Does your team have a specialist in this area? At the time of writing this book, one such topic is AI. Neural networks are used in all research areas. If this topic fascinates you and your team doesn't have a champion in this topic, then maybe this is an opportunity for you? The use of AI is becoming a standard, and you can be the one people come to with questions on the topic—and you'll certainly receive interesting collaboration invites from your teammates.

You can also develop in an area that isn't directly related to research. You can start attending events on collaboration between academia and industry and develop your knowledge in this area. Or maybe you can contribute to a group's wellbeing and be the person who's always organising group squash games or bringing in baked goods.

Being a champion isn't only about having the knowledge and skills, but also about wanting to pass them on. Therefore, if you want to be seen as such, you must take active steps in this direction. On the other side—it's really good to be expert and share your knowledge, but don't forget that your main focus is your Ph.D.

Takeaway points:

- **Each group has two layers: formal and informal.**
- **Formal structures can often overlap between research group, department and university.**
- **Informal structure includes "champions" in certain fields.**

1.5 Why Bother with Meetings?

Meeting After Meeting After Meeting

During your Ph.D., you will no doubt find yourself in many meetings. Meetings with your supervisor are just the beginning. Soon enough, you'll find yourself swamped by meetings with co-workers, group meetings, project meetings, departmental meetings, meetings related to teaching, meetings to plan the next meeting. That last one isn't even a joke—believe me, I've lived it quite a few times.

And you might wonder if they're all necessary. You might think some are just plain boring and useless and, more often than not, you'll be right. But there will also be a lot of important information for you at these meetings. Your attitude during them is also a crucial element in building your position and image in the group. Person showing engagement and professional attitude during the meeting is more likely to be involved in interesting projects and initiatives.

That's why you can't underestimate the role of meetings. And why you should take some of them seriously. It's why you should work out a proper meeting routine—and it starts before the meeting even begins.

Preparing for the Meeting

For a meeting to be worthwhile, everyone should turn up prepared. If you want to get ahead of the game, ask yourself these two questions before the meeting:

- **What answers do I need**—What should happen at the meeting to make it successful for you? What topics will be covered and what questions will you ask to get the answers you need? You could write down your arguments if you want to fight for your point.
- **How do I need to prepare**—Is there a reference material you need to read before the meeting? Review the agenda if available, have a look through the presentation from the last meeting etc.

I found this routine useful for both: one on one meetings and group meetings. But I believe that your routine would be different. Everyone must work out their own way of preparing notes and materials for the meeting. I use ReMarkable writing tablet. It allows me to group meetings by leading topics. In such way, before the meeting I can easily check the points discussed last time. This solution is most convenient for me and avoids the mess I have a natural talent for.

And there are plenty of other solutions I have tried before:

- Using one notebook for all meetings.
- Assigning meeting notes to individual binders depending on the meeting's subject.
- Maintaining several notebooks on meetings with different topics.
- Using dedicated software (Evernote, OneNote, Notion).

In time, and with some experimentation, you can work out the method that works best for you. The most important thing is that it's comfortable for you and doesn't require a lot of effort to follow the process. The more natural and painless the routine is, the greater the chance that it will stick.

Regardless of the equipment used, the fact of making notes allows you to use your meeting time effectively. It's also an expression of respect. By coming to the meeting with your notes and writing down new information, you give a clear signal that what your teammates are saying is important to you and you respect their time. I know a lecturer who refuses to meet with a student when they come to a meeting without a notebook. Don't let yourself to be that person!

How to Make Valuable Notes During the Meeting

You must find the way that works best for you. Too few notes might mean you can't recall what happened in the last meeting. Too many notes and you might fall behind and miss important thoughts. You must find that fine balance. It's not only about quantity, but the quality of what you write. How will those notes be useful for you in the future? From that perspective, it makes sense to distinguish different kinds of information. Most meeting notes contain two main types: reference information and action points.

Reference information—It is a piece of information that you may find useful in the future. Examples of reference information include:

- An interesting research paper worth reading.
- Information on how to calibrate the test equipment.
- Preliminary outline of the planned research project.

This type of material can remain in your notebook for use in the future. You can also copy it from meeting notes to your reference materials (which we covered in Sect. 2.5). It will be much more accessible to have this kind of information grouped by topic rather than dispersed in numerous notes from different kind of meetings grouped often by date.

Action points—All kinds of tasks that need to be handled by yourself. Examples include:

- Reading recommended papers.
- Talking about test equipment calibration with the maintenance team.
- Sending colleagues an email about the agreed preliminary project plan.

These points should be included in your task list. If you can't put them there right away, make sure to highlight your action points in meeting notes in a way you won't miss. This can be done, for example, by marking an asterisk in the margin or literally highlighting it with a bright, bold colour.

After the meeting, it's a good idea to look through your notes and recall the action points. You'll find that some can be addressed immediately, such as firing off an email summarising the meeting or making a phone call. The remaining action points should be logged in your calendar or task list, so you don't forget about them. Otherwise, let's be honest, you'll forget. I know I would.

Takeaway points:

- **Before the meeting, answer the questions why do I need this meeting, what answers do I need, and how can I get prepared?**
- **Your notebook is your best friend, and using it shows respect to other meeting attendees.**
- **During the meeting, organise the information into reference materials and action points.**

1.6 The End of Your First Year

You've Seen It All

It's here before you know it: the end of your first year as a Ph.D. student. You've made it to this very important milestone in your Ph.D. life Congratulations! If you're still before that milestone, then just know that the first twelve months are an amazing experience and a crucial period for your personal development. During this time, you will transform from a student to a researcher. You get to know your faculty and establish contacts with other researchers. You establish a rapport with your supervisor and start to understand your research problem in greater depth. You already know where to get the good gossip and how to talk to technical staff to get the job done.

If you're at the end of your first year, you probably feel more confident in your career direction thanks to the experience you've gained. Of course, if you have any doubts, this is a good time to review your plans for the future. But if you decide to continue your adventure, and I really hope you do, then you might need to convince others that you're ready to complete your Ph.D. with positive result. In many countries, there is a formal milestone at this stage, where you need to provide evidence that the work to date forms a solid platform for a Ph.D. level award.

Major Milestone

The exact procedures for evaluating a Ph.D. candidate and their project around the end of the first year may vary between countries, disciplines and institutions. However, in most cases, you'll be asked to write a report and there might be a formal viva when the examiner will ask you a range of questions based on your report.

This is a major milestone in the Ph.D. project process, which typically coincides with the stage of research where you conducted an in-depth literature review of the subject area, you undertook preparatory work for the body of the research work, and you have done a first exploratory use of the chosen research methodology or approach. The formal assessment process will determine whether your project has research potential at Ph.D. level, and to assess whether it's possible to complete your Ph.D. thesis during the standard period for the programme.

Regardless of the country where you do your Ph.D. or the formal process you follow, my advice is to focus on the following three points. This is what your examiner will be looking for:

- Do you understand the context of your work?
- Do you know what research methods you will use and why?
- Do you have a clear plan for completing your Ph.D.?

Let me elaborate on them in the next paragraph.

Three Ingredients

Firstly, without a thorough understanding of the context of your work, you obviously won't be able to successfully complete your Ph.D. In the last twelve months, you've read dozens, if not hundreds, of papers in your field. This is a time when you must prove that you understand how your work enriches existing knowledge. Which research question are you aiming to answer, and what is the rationale behind it? Who tried to address a similar topic and when? How does your research relate to already published results? If you successfully complete your Ph.D., what added value will your work bring and what problems will your results help to solve? These types of questions are to be expected in relation to the context of your work.

The second important point is the practical aspect, which relates to the research methods you want to use. You have to convince the examiner that you know exactly what tools you want to use to answer your research question. At this stage, you might be able to show the preliminary results of the pilot study. Rather than dwelling on the theoretical aspects of the methods, it's much better to present practical knowledge and concrete results. It isn't a problem if these are only preliminary results. What's important is to show that you're proactive and know what challenges await you. This will inspire confidence from the examiner and make them trust what you want to achieve.

The last point is to clearly explain the exact plan for the rest of your Ph.D. Based on a thorough understanding of the literature and knowledge of research methods, you should present a solid plan. Yes, a Gantt chart (see Sect. 1.2) may be helpful here, but it's more important that you can convincingly justify it. Make sure you have answers to questions that might crop up, such as:

- What are the main tasks you want to complete?
- What are the main milestones?
- What is the sequence of tasks?

The end-of-year-one evaluation is an important milestone in your Ph.D. project. Prepare yourself for it and it'll be a positive experience that will give you the confidence to sail through the rest of your journey.

Takeaway points:

- **The end-of-year-one evaluation is intended to identify whether you and your project have the potential for research at a Ph.D. level.**
- **The examiner will typically ask about the context of your work, methods used, and will look for a solid plan going forward.**
- **A well-written report will help with the formal viva.**

1.7 Battling the Inner Critic

You Are Not a Fraud

On the way to completing your Ph.D. course, you'll encounter many positive and encouraging experiences. But unfortunately, you'll face your fair share of moments of doubt, weariness, or procrastination. This is perfectly normal! Nobody is productive all the time and crises happen to everyone. If you're aware of them, you are prepared to face them head on.

It's possible that after months of hard work, you'll fall victim to what is known as "impostor syndrome". This is a term used to describe the mental state of a person who begins to doubt their own achievements. It may be accompanied by an overpowering inner fear that someone will discover that you've been faking it all along. You feel that you might suddenly become outclassed by your peers and that you've only made it to where you are now out of sheer luck.

This phenomenon has been the subject of many studies—one of the first ones worth looking at dates back to 1976. It was carried out at the State University of Georgia on a group of 150 professionally fulfilled women with doctorates in various fields who, despite many scientific successes, lived with the conviction that they actually owed their success to pure chance—certainly not their knowledge and competence.

Fighting Your Feelings

So how can you beat the imposter syndrome? You might be right—you are not the smartest person on Earth. And you can't just pretend to be smarter than you are. But that's where you can find your drive—you're determined to get there. Psychologists know this very well, that initially depressing thoughts can become mobilising, and people suffering from imposter syndrome can ultimately achieve success faster than those who have too much faith in their abilities.

What can you do to make yourself feel better and rebound from the destructive feeling of being an imposter?

- **Get objective**—Try separating your emotions from facts. This will help you to look at your real achievements and the progress in your work.
- **Catch your negative thoughts**—Moments of doubt are often provoked by specific events, words, experiences. Don't get caught up in the mental trap of negative thinking. Recognise the pattern of negative thoughts and break it by moving on straight away.
- **Find support**—Talk to your mentor (preferably a different person from your supervisor) or other students who are probably facing the same feelings as you.
- **Look into the future**—Imagine how great it will be to get that project finished and how well it will be received.
- **Look into the past**—Realise what a fantastic opportunity you have and feel gratitude for getting to that point. Looking into the past also allows to appreciate what you have achieved in the longer run. We often have a tendency to overestimate what one can do in the day, and to underestimate what one can do in the years.
- **Hug your failures, highlight your successes**—Try to find the positive aspect of any situation, even if it's a failure. In fact, every experience can be treated as a lesson that pushes you forward.
- **Treat yourself like your best friend**—We tend to be much more critical of ourselves than we are of others. Especially when we talk about the people we care about. Imagine that your best friend is in the same situation as yourself. What would you advise to her/him?

Don't be afraid of feeling down—that's what the brain does sometimes, so don't let it trick you. Even if you're a hard-working rationalist, you'll still have those moments of doubt. And that's absolutely fine!

Mental Health and Wellbeing

While the intellectual rigor of research is invigorating, the pressures of meeting deadlines, publishing, and navigating academic expectations can be overwhelming. It is vital to prioritise mental health as an integral part of the Ph.D. experience, recognising that a healthy mind is essential for producing impactful work and enjoying the journey.

Common stressors include the aforementioned feelings of inadequacy stemming from imposter syndrome, but also the isolating nature of research, difficulties maintaining work-life balance, uncertainties related to research outcomes and career prospects, and potential conflicts with peers or supervisors. These challenges, if unaddressed, can lead to burnout, anxiety, or depression, underscoring the importance of recognising and managing them early to maintain a healthy and productive Ph.D. experience.

Maintaining mental health during a Ph.D. requires proactive strategies to manage stress and build resilience. Developing a strong support network of peers, mentors, and family helps combat isolation, while effective time management ensures tasks are achievable and deadlines manageable. Prioritising self-care, such as regular exercise, a healthy diet, and mindfulness practices (see Chap. 7), supports physical and mental well-being. Setting boundaries between work and personal life is crucial for sustaining balance, and celebrating small milestones can boost motivation. Please take care of yourself! Parents are told in airplanes to put on their oxygen masks first, and then take care of their little ones. Likewise, you must take care of yourself to be in a position when you can make your Ph.D memorable. Last but not least, please do not be afraid to seek professional help when the feeling of anxiety or burnout becomes overwhelming.

Takeaway Points:

- **A Ph.D. is a massive challenge, and you aren't the only person struggling**.
- **Don't be afraid of feeling down at times—it's a natural part of the process**.
- **Maintaining mental health requires proactive strategies and taking care of yourself**.

1.8 Writing Thesis

What a Journey

Your Ph.D. will normally take three to four years to complete. On the surface, that sounds like a long time. And it is. But the good news is that at the end of that period, you'll be a different, better version of yourself. You'll develop in both a personal and professional capacity and, of course, you'll have a unique qualification in your pocket in the form of a Ph.D. degree.

I can assure you that when you look back, you'll fondly remember all the great moments from this fantastic journey. But to make this happen, you need to pass one of the most crucial, and most difficult milestones—get the thesis ready for submission. You want to put all your energy into that last stretch and finish on a high note. Finishing strong will build the momentum you need to move on quickly to your next gig. Author, speaker, and performance coach, Gary R. Blair, said it best: *"Many will start fast, few will finish strong"*.

Plan It Early and Keep Refining

The earlier you start drafting the table of contents for your thesis, the better. It will help you visualise the end product and align your daily work with it. The task of writing the thesis will become much more tangible and less intimidating. I recommend that the best time to develop an initial structure for your thesis is just after the end-of-year-one evaluation. At this point you should have a good idea of what you need to complete the project, and you should have a solid plan going forward. Most likely in the form of a Gantt chart. Of course, you will keep refining the table of contents as you go along, but with each update you will improve the flow of the thesis and your message will become clearer. It will also allow you to stay focused on what actually contributes to the thesis, and you will not lose time on "interesting" but distracting side projects.

A good starting point for planning the structure of your thesis is analysing of existing dissertations in your field. You can ask your supervisor to share a few examples with you or you can search the library sources. The final structure will depend on the type and variety of your results, but the main sections typically are:

- **Introduction**—Big picture of the study.
- **Literature review**—Sometimes contains the theory section.
- **Materials and methods**—Like in a research paper but more detailed.
- **Results**—Typically several chapters structuring your research output.
- **Discussion**—Critical assessment of your work.
- **Conclusions**—A summary of the key outcomes. Sometimes includes also the future work).

When planning your dissertation, think in terms of levels. The first level consists of the main sections, typically between six and ten, similar to the sections outlined above. The deeper level contains sub-sections. As you progress through your Ph.D., you should keep adding detail to your thesis plan and keep breaking the sections down into sub-sections. Mind-mapping is very helpful in this process as it allows you to visualise the structure of your thesis. You can even create a storyboard. This tells the "story" of the thesis in a small number of panels that mix text and pictures.

It's worth noting that some countries and institutions offer the option of "Ph.D. by publication". In this case, a degree is awarded to a candidate who has several published papers on related topics, which together form a portfolio of original work at the Ph.D. level. You could check with your supervisor to see if this option is available to you.

Hemingway Knew It

With a solid thesis plan in place, you will find it much easier to engage in the writing process. You will naturally want to add to the manuscript and fill your plan with awesome content. This is the right approach, as my experience shows that the best outcome is achieved when the process of crafting the thesis is spread across and extended in time. I would love to get this advice when I was writing up my thesis. It is some much easier to write in small chunks as you go along, than to lock yourself in the office for a few months to force daily grind. Believe me, I know what I am talking about. I relocated to family member's empty house to write up the thesis in one stretch of time working frantically towards the looming deadline. Not fun!

In Chap. 3, you will find a very specific advice on how to write a research paper. Some of that wisdom will be also applicable to writing a Ph.D. thesis, so go ahead and read it. However, before you run away, I have one more suggestion for you—think of your thesis as a collection of independent, self-contained sections. This approach has important positive consequences. Firstly, you don't have to write the sections in chronological order. Secondly, you don't have to waste energy thinking about the scale of your work and you can direct your focus on one section at the time. Again, bite-sized milestones rather than one, overwhelming task.

Finally, when it comes to writing, don't consider yourself wiser than Ernest Hemingway, who said it very directly *"The first draft of everything is shit"*. But the good news is that you will have plenty of time to review and amend your thesis if you follow my advice and start writing early. So have a look into your Ph.D. project plan. If you have the last couple of months assigned to "Writing up the thesis", change it to an ongoing task stretching from your end-of-year-one evaluation to the end of the project. Remember about milestones that will mobilize you to keep working on refining the thesis structure along the way.

Takeaway points:

- **Have the end game in mind, draft the structure of your thesis just after completing the end-of-year-one evaluation.**
- **Keep adding awesome content to your thesis as you go along instead of trying to write it in one go towards the end of your project.**
- **Hemingway famously said,** *"The first draft of everything is shit"*.

1.9 Preparing for the Viva

What Is Ph.D. Viva?

A viva, or viva voce, is defined by the Cambridge Dictionary as "a spoken exam for a university qualification". This precisely describes what it is: the final oral examination for your Ph.D. degree.

After sending your Ph.D. thesis to the examiners, you have to wait for your viva. You might find yourself with plenty of time to sit and twiddle your thumbs. The work has been done—the thesis has been submitted. All you have to do is prepare to meet with the examiners. Some Ph.D. candidates get nervous during this time and try to read whatever they can to prepare. This slight tension and adrenaline rush might help you prepare for your viva, but the paralyzing nerves are completely unnecessary.

I know that viva looks very differently across countries. Sometimes it is a public gathering where everyone is invited including your family members. In other places the viva is performed in front of the special committee or just in a room with the examiner and a chairperson. Every system is different. It's worth talking about it with your supervisor and someone who just went through the process. If the viva is public, it's good idea to take the chance of attending it. You can see it a couple of times—each time you will get more familiar with the process.

There is one thing which I assumed is common to all systems and countries—there are examiners who read your work in advance and want to discuss certain points during the viva. It's not a classic exam at all. You might indeed get some questions regarding your general knowledge in the relevant field. But majority of them will be related to your own study. Reviewers want to understand that you made an independent study, and you understand the meaning of your results. I could risk this statement—this is probably the only exam in your life, where you have better knowledge in the topic than your examiners!

Preparation

So you are the expert in your field, and if you thoroughly prepare there should be no surprises. Your main enemy will be your stress. Preparation will make you more confident and quickly release tension during the exam. Here's what I would recommend doing in the time between submitting the thesis and the viva.

Check the reviewers' profile—It's good to know who your reviewers are. Understanding their background will let you anticipate some questions and directions in which they could lead the discussion. As an excercise, you can prepare the list of possible questions you might expect from them.

Get ready for general questions—There is also a list of "classic" questions that everyone might expect during the viva. They usually refer to your perspective—what have you learned and what is your belief about the contribution of this work. Here are some examples:

- What's the originality of the work?
- What are the main research questions or hypothesis you have addressed?
- What's the impact of your work?
- What would you do differently if you start over today?
- What are possible directions for the future?
- What do you consider as the strongest point of your thesis?
- What are the biggest lessons you have learned?

Think of the toughest questions you could have—This exercise is mainly to let you release some stress. You will probably get many simple questions and a couple tougher ones. But let's get prepared to those you are really scared of. So make your blacklist of questions and leave it for a day or two. Then you can start responding to them with all the resources that are available to you. Prepare the answer and check it with your colleagues and supervisor. Once you have it—you are ready to the most pessimistic and dramatic version of the viva that might possibly happen. The real one will be much smoother, but the feeling of being ready is always helpful. As Jack Reacher says, *"Hope for the best, plan for the worst"*.

1.9 Preparing for the Viva

Read your work—If there is large timespan between submitting your work and the viva, give your thesis a quick read before the exam. Ideally, I would do this a week or two before the viva.

Prepare additional materials—If you think that some additional materials can become useful during the conversation, prepare them in advance. It's not certain that you will indeed use them, but sometimes just having them with you lets you release some stress.

Prepare some water—You will be speaking a lot! A bottle of fresh water will be very useful.

Relax and sleep well—It's really not necessary to go through all the books and papers on the day before the viva. You've got this—you've worked hard for a few years to get here. All you need to do is to put everything aside, relax and have a good sleep.

You might get asked to start the viva with a brief presentation. It will be an exciting challenge—you will need to present your entire work in just a few slides. And it is a known fact that that shorter presentations require more preparation. Winston Churchill described it this way *"If you want me to speak for two minutes, it will take me three weeks of preparation. If you want me to speak for thirty minutes, it will take me a week to prepare. If you want me to speak for an hour, I am ready now"*. You need to resist the urge to include all your results in the presentation. The goal of this exercise is to focus on the key aspects of your research and highlight its impact on the state of the art. You should plan for at least 2–3 full revisions of your slides, followed by a mock presentation to ensure everything is clear and effective. Sections 5.3 and 5.5 will help you in this process.

During the Viva

Very soon you will have your well-deserved Ph.D. You just need to respond well to the examiners' comments and questions. That's all what Ph.D. viva is about—conversation.

If possible, begin by thanking the reviewers for the time they've dedicated to reading and analysing your work. Then, transition into your opening lines, which you might have already practiced. Just like with any presentation, a strong opening can boost your confidence. I discuss this further in Sect. 5.5. Even if your viva doesn't include a slide deck presentation, you may still be asked to briefly summarise your work. Having your first sentences prepared will give you a moment to calm down.

Then, the questions will start flowing. Stay composed when answering them. The examiner isn't trying to trip you up; they're simply looking to explore your work further and will likely dig into the details they don't fully understand. They're genuinely curious about your perspective. If your viva leads to them asking for clarification, that's actually great news! It means you have the opportunity to explain more of your work and demonstrate your depth of knowledge.

If you don't understand the question, you can always ask for rephrasing it. Asking for repeating is usually not a good idea, as you might again have problems with understanding what you were asked about. If you are not sure you could always rephrase it yourself by saying "If I understand you correctly you are asking me about...".

When nervous moment comes, don't be afraid to ask for a moment to calm down. You are always fully entitled to take some water or to ask for a break or for a permission to go to a toilet. This will give you a moment to regain your composure and get back on track—on the path toward a successful and well-deserved Ph.D. degree.

Takeaway points:

- **Your final Ph.D. viva is a celebration of many years of your hard work, enjoy it!**
- **Be proactive and prepare for possible questions.**
- **Stay calm about questions, the examiners will be asking them to explore your work.**

1.10 Is There Life After Ph.D.?

Getting Out of a Bubble

It's very easy to get immersed in your Ph.D. You are fully focused on your research and spend most of time working on it. Depending on the discipline, that will involve long hours of coding, working with data, writing papers, reports or literally living your life in a lab.

I will tell you one thing—that's not how the other people live. And it's ok to have a different life for the time of your Ph.D. Someone is paying you to dive deep into an exciting subject and stay focused on it. You are in a bubble where only one thing matters—your research aim. That's for a reason. You can't discover something new and meaningful overnight. The only way to do that is to work hard, stay focused and non-distracted for a long time. And that's what your Ph.D. is for!

But one day you will get out of your bubble. I remember when I got out of mine. The day after viva I had a very strange feeling. It was a mixture of relief and anxiety linked to the fact that I had no idea about what to do with my day. Before it was simple: eat/code/sleep/repeat. It was really difficult to find myself in this new reality, where my research aim was no longer the main driver of my day.

To avoid such situation, it's best to prepare in advance. So, the question I am asking here is—"Do you know what you want to do after your Ph.D.?"

Available Options

You've probably chosen Ph.D. for a reason. My assumption is that you are a smart person who loves the stuff you do. That's good! There are a few professions for smart and passionate people.

Continue your academic career—If you've fallen in love with academic world, you can continue your journey. But you must keep in mind the following points:

1.10 Is There Life After Ph.D.?

- **Choice of the city**—You might want to change the place you live. If that's the case, it's worth considering moving to another city or country for a postdoc to immerse yourself in a new environment.
- **Choice of the institution type**—Do you consider university or different kind of institution? Most countries have national labs or academies of science. Those institutions are typically a little bit more competitive to get a job, but offer full dedication to research without the need of delivering teaching.
- **It will change**—It's worth noting that life of a postdoc if different than life of a Ph.D. You will never have the luxury of immersing yourself in one topic!
- **Consider the timeline**—Typically, it takes a few months to apply for a post-doc in academic institution. To avoid long break after your Ph.D. you should look for adequate position far in advance.

Industrial R&D—In certain disciplines you can find commercial R&D centres. Examples include AI, pharma, finance, energy, insurance. Such career path provides better financial conditions—your salary will be much higher. You will be also working on practical projects which have typically shorter route to real applications. All of this comes with the price of limited freedom. You need to bring the profit to your company. Therefore, your boss will have very exact idea of what should you focus on.

Work for non-profit—In certain disciplines you could find NGOs (Non-Governmental Organisation) that are doing research. Examples include public health, environmental science, education, human rights, economic development. NGOs typically don't offer salary advantage over university. But they are great place to work on projects that make impact to the society. Here are some examples of institutions offering that: World Bank, Oxfam, WWF, Bill & Melinda Gates Foundation (just as example of many foundations made by rich people who want to contribute to the world), or something which we are proud to be part of—"STEM Racing".

Consultancy—During your Ph.D. you've become an expert in a narrow field. Is it something that can be offered as a service? Such a career path gives you a full freedom on choosing what to work on. You will be likely involved in very practical projects. But you will need to build your own customer base—that's where Chap. 6 becomes useful.

Entrepreneurship—Ph.D. is all about being creative and finding new ways to solve problems. Entrepreneurship is not that much different—it's all about solving people's problems. Running a startup is an exciting way of doing something meaningful and on your own terms. But the life of a startuper could be very stressful. You will spend many hours speaking to the investors. Sometimes you will be living on a day-to-day basis. Not more than 10% of startups survive. But this is a high-risk, high-reward game! If you are among those great start-ups that can grow, your rewards will be much higher than anything you can think of in any of the abovementioned scenarios, at least in financial terms.

Journalism/popular science—If you particularly enjoyed "sharing the knowledge" part of your Ph.D. experience, then you might consider this option. With Ph.D. degree, you have all the credibility needed. You also understand the academic world and know many unwritten rules of that tribe. You can evaluate which research is worth noting to wider audience. All of that puts you in a good place to build bridges between society and academia.

How Would I Know What's Best for Me?

Time spent on Ph.D. feels like eternity, but in lifetime's perspective it's just one of many stages in your life. Do you know where will that stage lead you? Imagining yourself in 5–10 years from now might be a good exercise.

Another thing is to understand yourself. Do you know your superpowers? It takes a lot of self-awareness to understand that, but it's something worth exploring. The most important part of that process is to observe yourself during the day and try noticing when you feel best. Do you sometimes lose track of time? What actions make you feel great? Is there anything you have a natural talent for? Is there something that people say you are great at? In Sect. 2.7 I am also telling you about the "Strengths Finder" book which might be helpful.

I was lucky enough to meet John Scherer on my way. He is running phenomenal leadership workshops around the globe. One of his most powerful quotes is *"You don't need to change yourself. Ever. You need to come home to yourself. And that changes everything"*. It truly had a significant impact on me—instead of fighting with myself to be more like this or more like that, I focused on understanding what I really love doing. But how to know what is your passion and natural supertalents? That's on you!

The book "So Good They Can't Ignore You" offers an interesting perspective on the challenge of finding passion. The author, Cal Newport, critiques what he calls "the passion hypothesis", which is the belief that you just need to wait for a magical moment when you'll suddenly discover your passion. While this may happen for some people, Newport argues that it's not the best approach to simply wait for such a moment. He contends that, in most cases, inspiration doesn't strike out of nowhere. Instead, he advocates for the "craftsman mindset". It involves consistently working, paying attention to how you feel and perform, and focusing on improving the quality of what you do. Over time, this approach can help you discover your passion. So, rather than saying "follow your passion", Newport suggests you should "keep working and exploring yourself, and the passion will follow".

Takeaway points:

- **It's never too early to think about your career after Ph.D**.
- **Ph.D. is just a stage in your life. Do you know where you want to be in next 5–10 years**?
- **To better understand yourself you can train your self-awareness and observe when you are enjoying the process.**

Chapter 2
Understand and Organise the Context

> Understanding the state of the art and cultivating curious creativity will pave your way in the academic world.

2.1 Strategies for Managing Your References

Seasoned Researchers Have Their Own (Reference) Managers

When I began my Ph.D., I attended a meeting for new starters. It was led by an experienced colleague who was just finishing his own Ph.D. His message was: *"Each of you is working on something completely different, so it's difficult to provide precise guidance for your specific research project. However, I will give you a technical tip: when you return home tonight, install reference management software, and from tomorrow, make a habit of adding all relevant papers as you find them. This way, you'll steadily build your own scientific database and save yourself many days of frustration when it comes to writing your thesis and papers."*

To this day, I still think that was the best advice one could give to someone starting a journey in research, and I repeat it to all Ph.D. students I'm working with.

Over the years, I've conducted several workshops for researchers across a wide range of fields. In the first hour, I often ask participants about the strategies they use to organise the information they've gathered from research papers. Typically, around two-thirds of the group mention having a dedicated folder on their computer or in the cloud. Meanwhile, about one-third say they prefer keeping printed versions in folders, binders, or drawers.

Storing printed manuscripts or PDFs can be effective for up to about 100 papers. As the volume grows, we stop remembering where each new paper is and begin rebuilding bibliographies filled with the same pieces of research. We unintentionally lose days organising something we've already done. And when we search for a specific paper, it gets even worse. Do we actually remember things like the title, journal name, or publication year when we go back to find a paper? Probably not! Often, what sticks with us about a paper isn't its title or journal, but rather something like an insightful figure, the fact that it was presented at a conference in the USA, or one of its main conclusions. Can these kinds of cues guide how we structure our folders or directories? That's where the reference manager could can be helpful.

There are several popular reference management tools, including RefWorks, EndNote, Zotero, Mendeley, and BibTeX for LaTeX users. The particular software you choose is less important than committing to using one consistently. Pick a reference manager, start using it today, and systematically build your database. The earlier you begin, the sooner you'll have a substantial and valuable resource to support your research.

Knowledge is your capital and your tool. An accountant keeps their numbers well organised. A musician takes care of their instruments. An athlete regularly works on their physical form. If you want to achieve good results as a scientist, you should also take care of professional organisation of your knowledge.

And finally—if I didn't encourage you until now, I will need to ask Mr. Isaac Newton for help. One of the most famous quotes in science is coming from him—*"If I have seen further, it is by standing on the shoulders of giants"*. This metaphor is a clear tribute to previous researchers, emphasising that our work is merely a continuation and evolution of all the efforts made before us.

How Does This Work?

I used Mendeley, Zotero and JabRef to manage my references, and my database currently contains over 1,000 papers that caught my interest over years. When writing a paper or planning a study, I rely on this database to quickly access important, relevant information. It's efficient enough that I can simply enter a keyword into the search bar, and within moments, I have a curated list of related papers. Many entries are linked to PDF files stored on my disk, which I can view alongside my previous notes. Moreover, the Word integration allows me to edit my references on the go and cite items directly from my library. Tasks like formatting documents to match journal requirements and sorting references are handled automatically, saving both time and unnecessary frustration.

Have you ever experienced time when a reviewer asked you to add a new reference to your manuscript between citations number [3] and [4]? The paper has 34 references, which means that you have to renumber all items between 4 and 34. If you do this manually without making any mistake, please email me—you'll be the first person I know to pull that off.

What if you decide to submit your paper to another journal? It sounds easy enough, but after reading the formatting requirements, it turns out the bibliography needs to be in a different style and arranged alphabetically, instead of in order of appearance like the previous journal required. This manual reworking of the citation style can sometimes take several hours of work.

In each of these scenarios, a reference manager will handle the renumbering for you faster than it takes to read this sentence.

Which App to Choose?

Choosing a reference manager is like choosing a car. The truth is, every car serves its basic function well. Whether it's a BMW, Toyota, or Volvo, it will get you from A to B much faster than walking. If you ask a friend for their opinion, they'll likely recommend whatever car they own. The same applies to reference management. You can do it manually, just as you can walk from A to B on foot. However, if you use a professional, dedicated program, you'll be able to create your bibliography automatically in no time.

It doesn't matter which app it is, but you have to choose one eventually. Here is my list of factors that might influence your choice:

- **Compatibility with your colleagues**—just like you'd check with your partner before buying a car, it's worth discussing reference managers with your colleagues before making the final decision. If you all use the same software, it will be easier for your team to share the bibliography or collaborate on papers.
- **Integration with your work environment**—it's worth checking if the app works with your text editor (Word, LaTeX, Scrivener) and if it regularly releases new versions for your operating system.
- **Cost**—some of these apps are paid for and some are free. I recommend you visit the university library before making your choice, as some universities have already purchased access to a paid software. If not—you can always consider Zotero or JabRef which are community-driven.
- **Community**—if you are still unsure, you will probably google it. Some apps have strong communities and discussion forums. This is important—if you encounter an issue, you want to find a solution quickly. A common scenario you'll face is locating the correct citation style for a specific journal, particularly one run by a smaller publisher. To test how easily you can resolve such issues, try googling this problem and see how quickly you can find assistance.

Takeaway points:

- **A reference manager allows you to build your own database of useful papers and perform quick searches.**
- **Most apps work with your text editor and can generate bibliography automatically.**
- **It's recommended to have the same reference management software as your co-workers.**

2.2 Staying on Top of the Latest Literature (in a Smart Way)

State-Of-The-Art Changes Every Day

The number of scientific papers published in journals is growing rapidly. In fact, it's growing exponentially. This effect was first described by Derek J. De Solla Price in a series of lectures—"Little Science, Big Science"—published in 1963. He noted that the number of papers published in journals doubles every 15 years and this observation still holds today.

What does this mean for you? This means that today, dozens of new papers in areas related to your research are published every hour. Current publishing pace is close to hundred per hour per research discipline. Even if you sit in the library with the intention to read every new publication, you won't be able to keep up.

You need to find a way to track state-of-the-art, especially as responsibilities continue to pile up. Developing effective habits and leveraging tools that help you stay on top of the latest results in your field can make the difference.

Habits That Can Help You

Research paper folder—You could print intriguing papers right away and keep a hard copy in your bag. You never know when you'll be stuck on the airport or train station, or have some spare time between meetings. In such case, you always have a paper on hand which you can read or at least scan through.

Shared databases—It's always worth joining forces with others. If you maintain your publication database in the reference manager, consider sharing it with your teammates. This approach does require some cooperation and discipline, though. Every new paper added to the database should be introduced into the existing structure. Everyone must also use the same software.

Paper Club—This is an idea I noticed in different research groups. It's very effective but again requires good cooperation. "Paper Club" is a regular meeting organised by the research team. Everyone can participate under one condition: they will present a summary of an interesting research paper at the meeting.

Regular reviews—This habit lets you both learn and contribute to the scientific community. It's good to review papers for journals on a regular basis. This will make you a better author and also help you better understand the work done by other researchers. Section 6.8 will help you with the first review.

Tools That Can Help You

Reference manager newsletters—Some reference managers offer regular newsletters with paper suggestions. Adding another newsletter to your already overloaded mailbox might not sound too sexy, but if you give it a second thought this might be actually the one which is indeed useful. Knowing your library, the reference manager has a very good understanding of your research interests. On this basis, it suggests new papers and the recommendations are quite accurate. It's a smooth way to find interesting papers without much effort.

Search engine alerts—Search engine alerts work very similarly. If you regularly use Google, it knows exactly what your scientific interests are. It's well worth taking advantage of this. To do so, just activate the Google Scholar alert with paper suggestions based on your search history.

2.3 Can State-of-the-Art Be Two-Dimensional?

RSS feeds—You're probably familiar with RSS feeds. It's worth checking out if you can use it in this context. Perhaps there's an RSS feed that will allow you to receive fresh information about scientific reports from your industry? Scientific organisations and publishers often have RSS feeds that you can subscribe to.

Save for later—There's a wide range of tools allowing you to save an interesting paper to read later. A popular one is "Instapaper". Many browsers offer "save for later" option. You can also use dedicated folder for that purpose. It's no different from printing papers for the future. Only this time you don't have to carry the file with you. In Sect. 2.4 I will talk about strategies for storing data and information. You might be looking for the ultimate tool that does all of that and there are many options to choose from: Notion, OneNonte, Evernote, Google Keep or Fusebase.

Scientists' profiles—Some scientists are very active on social media. It is worth following a person who is a recognised authority in your field. This way, you can stay up to date with their own research. Chapter 6 is where I describe all kinds of online profiles researchers can have.

Takeaway points:

- **The number of research papers in your discipline doubles every 15 years.**
- **Create habits to track new research, e.g. review for journals or launch "Paper Club".**
- **AI can help you: sign up to alerts and newsletters.**

2.3 Can State-of-the-Art Be Two-Dimensional?

A List of Papers Is Not Enough

Now you understand why organising and reviewing your literature is such a crucial starting point for your Ph.D. I've also explained why a reference manager is an invaluable tool for this task. Next, let's explore an additional method that can be especially helpful when you're taking your first steps in research. Many Ph.D. students worry about managing the vast range of papers, books, and sources they'll encounter. While storing them, as I've discussed in previous sections, is important, it's often not enough. A more structured approach may be necessary to truly stay on top of everything.

When you begin building your literature review, it might feel like stepping into The Matrix. There's that buzz of excitement as you enter academic realms previously unknown to you. But at the same time, there's a looming sense of dread—each paper approaches problem differently making it nearly impossible to compare all of them. But what if there's a way to control that chaos and make every piece of literature work together in harmony?

If you take the blue pill and skip this section, the story ends. You'll wake up in your own bed and carry on as you were. But take the red pill, and you'll discover the secret of the matrix review method.

Two-Dimensional Literature Review

The matrix review method involves creating a matrix that organises key information from the papers you've read and can refer to in your future work. Here are the main advantages of this approach:

- It allows you to quickly compare papers in terms of research methodology, publication time, key conclusions, or another other criteria important for your project.
- It highlights differences and similarities between papers, which is particularly valuable when many publications are related to similar topic, and you have trouble remembering the subtle differences in the arguments of individual authors.
- It presents a synthesis of the most important conclusions from a given publication, not its full summary.
- It provides a quick way to refresh your memory about the content of the sources you've collected (especially helpful when you have many).
- It allows you to view each source both in the context of other works and on its own, giving you a clearer understanding of how it fits into your research landscape.

The concept of the matrix review is highly flexible, and ultimately, you are the one who decides which information from each paper is most important to include. The content of your matrix will likely vary depending on your discipline and the specific information you are looking for in the publications you review.

How to Start?

Create a spreadsheet, which will allow for easy expansion in any direction. This format also makes it simple to share your matrix with peers, facilitating collaboration and feedback.

The example below shows an initial review matrix for the topic of Physics-Informed Machine Learning. The first five rows include basic information and links to the relevant files. Following these, a brief summary of key findings is provided, enabling a quick recall of the content. The subsequent rows address specific issues covered in the publication.

2.3 Can State-of-the-Art Be Two-Dimensional?

File location	...	...	...	...	...
Title	Physics-informed machine learning: case studies for weather and climate modelling	Physics-informed machine learning	Physics-informed machine learning and its structural integrity applications: state of the art	Scientific Machine Learning Through Physics–Informed Neural Networks: Where we are and What's Next	Lift & Learn: Physics-informed machine learning for large-scale nonlinear dynamical systems
Authors	K. Kashinath et al	G. E. Karniadakis et al	S-P. Zhu et al	S. Cuomo et al	Elizabeth Qian, Boris Kramer, Benjamin Peherstorfer, Karen Willcox
Publication year	2021	2021	2023	2022	2020
Discipline	Atmospheric Science, Meteorology	General Machine Learning and Computational Science	Mechanical Engineering, Structural Integrity	Scientific Computing, Applied Mathematics	Applied Mathematics, Mechanical Engineering, Dynamical Systems
Key findings	Achievements in physical consistency, accuracy, and generalization in weather and climate models	Connections between classical numerical methods and ML	PIML improves prediction accuracy and consistency with prior knowledge in structural integrity applications	PIML shows promise in scientific computing, but theoretical issues remain unresolved	Lift & Learn models capture system physics accurately and are robust to input changes, outperforming traditional methods
Programming language	Python	Python	Python	Python	Python, MATLAB
Network configuration	5 layers, 100 neurons per layer	5–10 layers, varying neurons per layer	5 hidden layers, 250 neurons per layer	3–10 layers, 32–250 neurons per layer	4 layers, 100 neurons per layer
Future directions for PIML	Addressing challenges in developing robust PIML models and improving generalizability	Need for further research on theoretical issues and applications	Identifying critical research directions for PIML in structural integrity	Exploration of causal representation learning and hybrid problems in PIML	Investigating theoretical questions of quadratic liftings and automated discovery of lifted coordinates

And that's how you do it! After introducing the first few papers, it's a good idea to present the matrix to your supervisor and ask for their feedback. I believe they will appreciate the effort you've put into organising the information and will be happy to support this process.

Creating the matrix may present several challenges. One common pitfall is the temptation to include too much information, such as entire quotes from the papers. Try to resist this urge and focus only on the key points that will help you quickly recall the content after several months. As you add more entries, you'll likely notice that some points in the key findings section are repetitive—these can be good candidates for separate rows. Additionally, you don't need to fill every cell. For example, if a paper doesn't discuss the future application of PIML, simply note "not discussed."

In the early stages, you may feel uncertain about the process, especially when the row titles aren't perfectly defined. However, don't put too much pressure on yourself! The most important step is simply to start organising your sources. Over time, you can adjust the matrix as your needs evolve, adding new rows or removing unnecessary ones. Remember, the matrix is a tool to help you navigate and manage your research—its primary purpose is functionality, not aesthetics.

Takeaway points:

- **A literature matrix is a table that lists papers alongside key features or outcomes.**
- **This method is particularly powerful for those starting a scientific career or entering a new field.**
- **Don't get too fixated on the matrix structure—it will evolve as you add more entries.**

2.4 Tips for Organising Your Data

You Will Generate a Lot of Data

During your scientific adventure, you will undoubtedly generate, process, and collect a wide variety of information. It will come to you from various angles—sometimes as loose inspiration, sometimes as structured notes from a book you read, and sometimes as a new dataset for your research. You need an approach that will make it easier for you to not only collect, but also organise your charts, notes, ideas, and data.

In the past, such a function used to be performed by a good old notebook, but today it's often insufficient to cope with the amount of incoming information and its various forms. Just think about them all: research papers, conversation notes, charts, emails, pictures, films, voice recordings, datasets—the list goes on. Let's collectively call these "reference materials".

It's worth explaining here what you should treat as a reference material. My definition is taken from "Getting Things Done" book by David Allen, which describes the reference material as *"information or material of a substantive nature that may prove helpful to you in the undefined future"*. A reference material is therefore: an interesting slide from a presentation, a fragment of a paper, graph, conclusions from a certain stage of research, laboratory procedure—you get the idea.

What about materials that are strictly connected to a particular moment in time? Take, for example, a "call for papers" email inviting you to submit an abstract for a conference. This email contains both a task (to write an abstract) and the material (the abstract template attached). This doesn't qualify as "reference material," which is defined as "information or material of a substantive nature that may prove helpful to you in the undefined future". The issue is that the template won't be needed in the undefined future; it will be needed at the specific moment when you're writing the abstract. For this reason, you should avoid adding it to your reference materials library, as that could lead to an overload. This kind of information should be handled differently, and I will refer to it as "Time-related reference material."

It is worth organising reference materials at the early stage where you are just collecting them. In the future, when you need a particular information, it will save you a lot of time (and sometimes frustration). Imagine you're putting all your reference materials in a drawer without sorting them. I guarantee you that after a few weeks, you'll run out of space, and it'll make you anxious just thinking about finding something there. You'll probably then decide to store new reference materials in the next drawer, then in the closet, and so on. I remember visiting one professor's office where the whole floor was occupied by a 1-m layer of papers, among which there was only a path to the desk and the closet. I felt like I was in a trench! It all could have started from innocently putting one sheet of paper aside. Fortunately, this professor was a person of incredible knowledge and mind—he probably didn't really need all these sources for effective work. However, one can guess what incredible things he could have achieved if he had properly organised materials at hand (and if he had the opportunity to freely walk around his office).

Tools for the Job

The tool you choose should be your most faithful assistant. It should be with you all the time and let you store your materials with minimal effort. If it takes a lot of effort to add a new reference material, it is only matter of time, when you skip adding some items and the workflow becomes leaky. So, let's have a look on your work routine. Think about the typical materials you want to store—loose thoughts or hard data? What devices accompany you during your work—tablet, phone, or computer? And how has it looked like so far?

It's possible that you'll use more than one tool for different types of reference materials. You can store experiment data in the cloud, notes from meetings in a notebook (or tablet), and the most interesting papers in the reference manager. It is important not to duplicate functions. If you intend to use more than one solution for storing the same kind of information, you will be always unsure where this particular item is archived. Be also prepared to change your workflow. It will be a lifelong process of reviewing your current setup and adjusting to changing character of your work.

Up until now, I've been offering general advice, which may feel somewhat abstract. To make things more practical, I'll share the tools I use. I'm not claiming that this is the perfect combination—indeed, it may not be suitable for everyone—but here's a practical example of how this might work:

- **Research papers**—I store all my research papers in the reference manager I'm currently using (they all serve the purpose).
- **Digital notes**—Materials taken from websites, emails, and files are stored in FuseBase. It's very convenient, as I can easily copy and paste whatever I think will be useful. Information is grouped into folders that refer to different issues, projects, or ideas. Similar aps include OneNote, Notion, Evernote, and Google Keep.
- **Handwritten notes**—Since I love taking handwritten notes, I always have my ReMarkable tablet with me. I use it for meeting notes and brainstorming sessions. This tablet synchronises with the cloud and allows me to organise notes into notebooks linked to specific issues, projects, or ideas. You can achieve similar functionality with a regular tablet or even an old-school paper notebook (which has the priviledge of eternal battery life).
- **Materials in shared apps/documents**—In some contexts, you need to collaborate with others on your reference materials. For example, you might be involved in a separate research project alongside your Ph.D. or have joined a conference organising committee. Since you're working with others, your choice is limited to environments everyone in the group is familiar with. This often narrows down the options to: Google, Microsoft, Apple, and possibly Slack, Discord or Dropbox. I personally use Microsoft 365 Groups, as it's available through my university account. By creating an Outlook group, I get a shared email that distributes messages to all members and a shared drive for everyone to upload reference materials. That's all I need, but if you require more features, you can also use other tools automatically included in Microsoft Groups, such as Teams channels, OneNote, and Microsoft Planner.
- **Time-related reference materials**—Since this type of material is linked to specific tasks, I incorporate it into my planning process. I manage my personal tasks using Amazing Marvin, which is highly flexible and lets me experiment with various planning strategies. One strategy I use is to attach reference materials to a specific task, making them accessible when working on that task. Similar apps include Microsoft Planner, Google Tasks, ClickUp, Trello, Asana, and Remember the Milk. If you rely on classic Google or Outlook calendars, you can also add notes and attach files to events.

The Importance of Inbox

You won't always be able to decide the location of reference material at the moment you receive it. Sometimes, you simply won't have the time. Imagine, for instance, that during a meeting, your supervisor recommends an interesting paper. You don't want to disconnect from the discussion to add the reference to Mendeley. What you need in this case is a place where you can quickly jot down a note about the paper and organise it later. This is where your inbox comes in—a space where all reference materials can go before you sort them and transfer them to the appropriate app.

To make your inbox work effectively, you should consider the following elements:

- **Easy input**—It should be quick and seamless to add any type of material to the inbox. Sometimes you'll be adding materials in the brief moments between your commitments, so speed is essential.
- **Easy output**—It should be just as easy to move materials from the inbox to their permanent storage location. If this process is long or painful, you'll eventually stop doing it.
- **Digital or analog**—The original "Getting Things Done" methodology suggests an inbox in the form of a physical space (like a tray or folder). However, I believe that a digital inbox is more appropriate, as it's more versatile and accessible.
- **Regular revision**—An inbox that's too full will quickly become counterproductive. Setting aside regular time—perhaps weekly sessions—to organise materials from your inbox will ensure the whole system works smoothly.

Even though organising knowledge and data may seem like a daunting, titanic task at the beginning, remember that a well-designed system will serve you for years and significantly speed up the process of writing papers, preparing presentations, applying for grants, or conducting research in general. If you're intimidated by the scope of the task, break it down into smaller steps. Spend just 10 min a day organising your desktop, folder by folder. After only a few days, you'll appreciate the sense of control that comes with having everything in its place.

Keeping your reference materials organised not only helps you work much more effectively, but also reduces stress. This approach is especially beneficial for people who tend to unintentionally create chaos. I'm speaking from personal experience—without the discipline to organise my data, I would be completely lost.

Takeaway points:

- **Reference materials include information of a substantive nature that may be helpful in the future.**
- **Think of a strategy and tools for storing reference materials that works for you.**
- **Secure inbox space, a place where reference materials will go to be sorted.**

2.5 Natural Planning Method

What Is Natural Planning?

"*If you fail to plan, you plan to fail*". That's what Benjamin Franklin said, and I believe—he also meant your Ph.D. When planning, it's important to take the right sequence of actions. There are many publications and sources on this subject. Here, I'll refer again to "Getting Things Done" by David Allen. In one of the chapters, Allen defines the sequence of actions in an effective planning process. He named them "natural planning method":

- Defining objectives and principles.
- Presenting a vision of the final results.
- Brainstorming.
- Tidying up.
- Identifying the next steps.

Let me give you an example of two different ways: one using a reactive planning model and one using Allen's natural planning model.

Example of Reactive Planning

You enter the lab. Considering the available reagents and equipment, you plan the next few working days (identification of the next activities). As planned, you perform a series of experiments using the procedure you know well. After three months of work, you collect all the data. After processing the results, you attempt to arrange them in a logical sequence (tidying up). It appears to not be that easy, and you have some problems interpreting the results. So, you ask your colleagues for help. You organise a meeting where you discuss possible conclusions from the results (brainstorming). During the discussion, an idea is born to use the results to write a paper about the use of a new reagent to increase resistance to disease (presentation of a vision of the final results). You also decide to ask a well-known professor and head of a large lab in this field for advice. After hearing your arguments, the professor asks you this question—why you want to do this (defining objectives and principles)?

Throughout this example, you've been reactive to the situation, responding to your environment. Such an approach is often driven by the belief that the sheer volume of tasks leaves no time for reflection or planning. You must move on to action immediately! In the reactive model, we do everything in the opposite order to the natural planning model: we start with the current activities, then organise them, then give them a meaning. As a result, defining our objectives is the last stage and it only comes out of necessity. I see this situation happening very often, even to experienced researchers. Everyone gets sometimes trapped into reactive model.

Example of Natural Planning

In one of your early discussions, your supervisor brings up the topic of the core values and goals you wish to achieve during your Ph.D. After several meetings, you both agree that quality is a top priority, and you want to ensure that your work is credible and reliable. Looking beyond your Ph.D., you aim to become "a leader in researching disease X resistance." This is a challenging but exciting goal, and you're passionate about the topic with a keen attention to detail (defining objectives and principles). The next step is to figure out how to make this vision a reality. You begin by researching how established leaders in the field operate. On this basis, you formulate measurable goals to be achieved over the next five years. These goals include researching the X model, publishing in the X journal, presenting at the X conference, establishing a partnership with company X, and co-developing a prototype (presentation of a vision of final results). You then brainstorm with your supervisor about the best ways to reach these goals. You discuss potential experiments, the reagents you'll need, the research methods to use, and where to secure funding (brainstorming). From this brainstorming session, you select the most promising ideas and begin planning the steps required to move forward (tidying up). You ensure that for each objective, you identify the resources needed and the immediate next actions to take (identification of the next activities).

The natural planning model begins by defining your aims and principles. When these areas are clearly defined, decision-making becomes considerably more straightforward. Before taking any action, you can ask yourself whether it brings you closer to achieving your goal. As Roy E. Disney famously said, *"When values are clear, decisions are easy"*. By aligning your actions with your core values, you ensure that every step you take moves you toward your intended destination.

The natural planning model fundamentally emphasises the importance of having a clear and defined goal. David Allen states that starting a meeting with an unclear prompt, like "Does anyone have any ideas?" probably won't lead to productive discussions. Instead, the conversation should start with a clear objective in mind: "What do we want to achieve?" Idea generation becomes most effective when everyone is aligned with the aims and principles and understands the desired results.

Takeaway points:

- **Reactive approach means your actions are reacting to circumstances (you have little control over them).**
- **Natural planning means your actions result from your aims (you have full control over them).**
- **To make sure you follow the right approach, compare your actions with the five points of natural planning.**

2.6 Being Creative in Your Research

Creativity—Humas' Lost Talent

Do you think that creativity can be helpful for your Ph.D.? Oh yes! Let's just follow another quote from Albert Einstein *"The true sign of intelligence is not knowledge, but imagination"*.

Do you think that creativity can be learned? Oh yes! In fact—most of us need to re-gain that skill.

It's difficult to admit, but most of us were probably much more creative in childhood than we are today. I learned this from the book "How to rise successful people", by Esther Wojcicki, named as Godmother of Silicon Valley. She is a famous teacher known for running a super-creative journalism class in Palo Alto. Achievements of her three daughters speak for themselves, they reached positions such as: professor at the University of California San Francisco, CEO of YouTube, co-founder of the company 23andMe.

In the book Esther wrote *"In one study, a test based on NASA's recruiting process for engineers and rocket scientists was used to measure creativity and innovative thinking in small children. At age five, 98 percent of the kids had genius-level imaginative abilities. But at age ten, only 30 percent of the children fell into that category. Want to guess how many adults maintain their creative thinking skills after making it through our educational system? Just 2 percent."*

Teaching can be a powerful force for positive change, but in many parts of the world, much remains to be done to harness its full impact. Often education focuses on memorising "the right solution," which is valuable to some extent. However, if we concentrate solely on learning and repeating these right answers, we miss the opportunity to foster creativity. When we encounter situations where the right answer isn't immediately clear, we may find ourselves unsure of how to approach the problem or think critically. Success depends on striking a balance between mastering proven solutions and embracing the freedom to explore and innovate.

Finding Smart Ways for Solving Problems

You may be familiar with the feeling of grappling with a research problem that just won't yield a solution. The results contradict your hypothesis or are internally inconsistent. You can't seem to explain the phenomenon, or your apparatus isn't functioning as you expected. Despite numerous attempts, you're not achieving the breakthrough you're hoping for.

Instead of continuing down the same path, why not try approaching the problem differently?

Change your environment—If you're used to solving equations in your office, consider relocating to a new "box." Sometimes a change of scenery, like stepping into a park or a café, can help shift your perspective. You might also find that working at a different time of day brings new insights.

Search for analogies—Perhaps a similar problem exists in a completely different research area. Look into fields like medicine or military research, where there are often large budgets and complex challenges that could offer insight. It's also worth considering whether nature itself has already solved the problem—after all, evolution has had a long time to work out optimized solutions.

Forget your experience—While your past experience can certainly help, it can also limit your creativity. Your mind tends to look for shortcuts, recalling how you solved similar problems before. As Paul Arden writes in his book "It's Not How Good You Are, It's How Good You Want To Be"—"*Experience is built from solutions to old situations and problems. The old situations are probably different from the present ones, so that old solutions will have to be bent to fit new problems (and possibly fit badly)*".

Involve students—If you're working with students, consider turning your problem into a project for them. Since they don't know "the right way" to solve it, they might come up with a solution that you would never have thought of—and it could work even better.

Do the sprint—For big, complex problems that require deep focus, consider organising a dedicated sprint session. This could be just for you, or for a group of colleagues (the more diverse the group, the better). Cut off all distractions for a few hours and concentrate solely on solving the problem.

2.6 Being Creative in Your Research

Creative notebook—Treat your ideas with respect and don't let them slip away. Even if you're busy with a paper deadline and can't explore a new study idea right now, make sure to record it somewhere. A dedicated notebook or a space in your reference materials can serve as a "backburner" for future ideas, ensuring they aren't forgotten.

Piece by piece—Another effective approach for solving large problems is to break them down into smaller, manageable tasks. Schedule several shorter sessions, around 30 to 60 min each. The first session should focus on defining how to divide the big problem into smaller pieces. In each subsequent session, devote yourself entirely to one of the mini-tasks you've outlined.

Use the magic moment—There are certain times of day when your mind naturally disconnects from routine activities, and your creativity flows more freely. This is the perfect time for a dedicated internal brainstorm session. Why do so many ideas come to you in the shower, while driving, or while running? You can make better use of this hidden potential by consciously deciding to focus on a particular problem during these moments. This could even be a reason to enjoy hiking without feeling guilty about not working on your Ph.D.—because your mind is still working, just in a different way.

Talk to people outside your lab—Broaden your perspective by engaging with individuals from different backgrounds. Some experts attribute the success of Oxford and Cambridge universities to their college structure, where students from various disciplines live and eat together, fostering the exchange of diverse scientific perspectives. If your institution is more discipline-based, seek out interdisciplinary conferences or events to connect with people beyond your immediate field.

Use dedicated methodologies—There are frameworks to guide the idea generation process, such as TRIZ, Design Thinking, and LEGO Serious Play.

Sleep well—The saying "I need to sleep on it" holds true: scientific research shows that good sleep habits support creativity. We'll explore this in more detail in Sect. 7.2.

Go to courses—Personal development courses, whether focused on creativity or new skills, can offer fresh perspectives and inspire different ways of thinking. More on this in Sect. 2.7.

Change your environment—An internship or study visit places you in a new setting with different people, helping you learn how others solve similar problems. I'll delve deeper into this in Sect. 6.9.

Creativity Is a Muscle

You can train your creativity just like a muscle. It's about getting back into shape, much like after the festive season. The key is in finding habits that encourage your mind to work in new ways. Just as anything that stimulates your muscles promotes growth, anything that pushes your mind to explore new neural pathways will enhance your creativity.

It's difficult to generate new ideas if you're stuck in the same routine—working in the same way, looking at the same view through your window, talking to the same people, and discussing only a limited range of topics. This is a perfect recipe for stifling creativity. Let breaking your routine and stepping out of your comfort zone become a habit in itself.

Takeaway points:

- **We've been much more creative as kids.**
- **Experience can be valuable, but it may also hinder fresh, new ideas.**
- **Breaking the routine could let you re-gain some of the lost creativity.**

2.7 Investing in Yourself

Henry Ford Gives Us a Recipe for How to Stay Young

Henry Ford once said: *"Anyone who stops learning is old, whether at 20 or 80. Anyone who keeps learning stays young. The greatest thing in life is to keep your mind young."*

Undoubtedly, you are driven by the ambition to become a leader in your field. And rightly so; if you want to be recognised as an expert, you can't stop your education at the master's level. Any time is good for further education. Being in a university environment opens up opportunities which are often not available to others. That's why it's worth looking for groups and organisations on campus that allow you to develop yourself: open lectures, meetings, interest groups, career development centre, and more. It pays back to be active and curious.

The first step is to assess your current needs by evaluating your strengths and weaknesses. This self-assessment will help you identify the areas where improvement will be most beneficial, allowing you to focus your efforts where they'll have the greatest impact on your progress.

Start by consulting with your friends, colleagues, and supervisor. Their honest feedback about your strengths and weaknesses can provide valuable insight. You can also consider taking a personality or skills test to gain a deeper understanding of yourself. Books like "Strengths Finder 2.0" by Tom Rath can be helpful in this process. Rath emphasises that focusing on developing your strengths rather than trying to improve weaknesses can lead to significantly better results. The challenge, however, is that we often aren't fully aware of our own strengths. A good personality assessment can reveal aspects of your character that you may not have realised, enabling you to leverage these strengths for greater success.

The Three Percent Rule

Before we move on, let me tell you about the three percent rule. It's a rule defined by motivational speaker Brian Tracy. In his book, he wrote: *"Here is a rule that will guarantee your success—and possibly make you rich: invest three percent of your income back into yourself."*

Many successful companies follow the rule of investing at least 3% of their profits into employee development. Why this specific amount? It's small enough to be included in the financial plan without causing significant strain on the company's budget, yet it's substantial enough to produce noticeable, positive effects relatively quickly. This balance ensures that companies can continuously improve their employees without overburdening their financial resources.

Regardless of whether your organisation follows the 3% rule or not, I recommend applying a similar approach to your personal development. Some training and materials crucial for your growth may not be available through the university, especially if they aren't directly tied to your current project. However, don't hesitate to invest in yourself. There are numerous valuable books, courses, and training opportunities that can provide fresh insights and inspiration. The key is to select them wisely, ensuring you make the most of your investment.

Finding the Development Opportunity

Here are some questions you might ask when searching for a perfect course or workshop.

Soft or hard skills?—This is the most traditional way of dividing the types of skills. Soft skills are personal traits that can be developed and contribute to your growth, such as communication, teamwork, and adaptability. A hard skill is a specific competence, such as the ability to use a particular piece of experimental equipment or software. You should try to balance and pay attention to both.

Group or individual?—Individual training is ideal if you want rapid improvement in a very specific area with the guidance of a dedicated tutor or coach. On the other hand, group training is often more affordable—and that doesn't necessarily make it less effective. In fact, when the course is a workshop-style, group settings can be especially beneficial, as they foster collaboration and diverse perspectives.

Online or in-person?—This largely depends on the skill you wish to develop. For example, a data analysis course can be effectively delivered online, often providing great value for money. However, the downside is that online courses require a higher level of self-discipline, as you'll need to stay motivated and commit sufficient focus time to complete the course effectively.

Who is running it?—Knowing who is behind the course can make a huge difference in your decision to enrol. Learning from a leading expert in the field is always a valuable opportunity. If you're unfamiliar with the tutor, take some time to research their background. Look for any materials they've published, such as papers, books, or lecture notes. Additionally, webinars or videos of past lectures can give you insight into their teaching style and whether it aligns with your learning preferences.

Who attends it?—Consider the people who will be on the course with you. Some courses offer alumni status and the opportunity to join a new network, which will last long after the course finishes.

Has someone recommended it?—Did you discover this on your own, or was it recommended to you? Remember, everyone needs different support at different times, so not every recommendation will be relevant to you. Instead of asking "Was the course good?" ask "What was it like?"

Takeaway points:

- **Never stop learning—you deserve more.**
- **Invest at least 3% of your time and income into personal development.**
- **Developing your strong points is more effective than fighting your weaknesses.**

2.8 Racing with Technology

Technology Is Advancing at a Faster Pace Than Ever Before

Consider how much has changed in just a few decades—or even a few years. Technology is no longer solely the domain of engineers; its influence now spans every discipline, from philosophy, law to literature and social sciences.

Something we don't always grasp is the nature of this acceleration. Our smartphones are billions of times faster than the first computers, yet they cost far less than computers did in the 1980s. As it turns out, technological development follows an exponential—not linear—trajectory. Let'slook at a simple example to compare the two concepts:

- **Linear**—If you take 30 steps of one metre each, you will be 30 metres away.
- **Exponential**—If you take 30 exponential steps starting at one metre—doubling the distance each time—you would cover 536,870,912 metres, enough to travel more than 13 times around the Earth's circumference.

The key to predicting the future of technology is understanding the exponential curve. At first, the increments appear to grow slowly, almost like a linear sequence: 1, 2, 4, and so on. Within just a few more steps, it skyrockets—surpassing linear growth by several orders of magnitude. Also, half of the total exponential growth takes place in the last step.

The phenomenon where progress accelerates over time is described by futurologist Ray Kurzweil as the "law of accelerated returns". This principle suggests that more advanced societies possess greater capacity for rapid development compared to less developed ones. For instance, 19th-century civilization had more knowledge and superior technology than that of the 15th century, enabling it to achieve significantly greater progress within the same span of time.

In the evolution of technology, acceleration is the norm. One of the most striking examples of this consistent exponential progress is the Moore's Law. To be precise, it is an observation, but is indeed widely named as the law. It states that the computational power that can be packed into a given size of an integrated circuit doubles approximately every two years. This principle, articulated by Gordon Moore in 1965, has remarkably held true for decades, driving extraordinary advancements in computing technology and its applications. Every few years, researchers announce that Moore's Law will no longer hold in the near future, only for a new technological breakthrough to emerge shortly thereafter, ensuring its continuation.

2.8 Racing with Technology

We can assume therefore that the exponential growth of data processing will continue in the foreseeable future. Ray Kurzweil notes that although the current technology cycle is driven by integrated circuits, it will not end when we exhaust their potential. After all, we already have quantum computers on the horizon, and as exponential growth continues, we can expect further increase in computing power over the coming decades.

What Does It Mean for My Research?

To effectively anticipate the future of your research field, it's essential to understand the implications of exponential growth. The most intuitive—though flawed—way of predicting progress is to think linearly. When considering the next 30 years, our instinct is often to look back at the progress of the past 30 years and expect similar scale of change in the future. This approach aligns with how our intuition is wired, and it has worked well in many situations throughout history. For previous generations, exponential growth didn't significantly exceed linear trends over timeframes close to a single human lifetime.

However, we have now reached a point where linear intuition only mirrors exponential reality over very short timespans. Over longer periods, this linear mindset leads to substantial underestimations—just consider the difference between 30 and 536,870,912. Those who recognise this need to override their natural instincts with scientifically grounded observations. It's similar to accepting that the Earth is round (or more precisely, an ellipsoid) even though our intuition suggests it's flat. Instead of extrapolating linearly from the past, predicting the next 30 years requires acknowledging the current pace of change and accelerating it exponentially.

We often hear about the "recent rise of AI," as though it appeared from nowhere. However, the history of AI stretches back to Alan Turing's work in the 1950s. It has taken decades for this technology to reach its current state. Like all exponential growth, it began slowly in its early years. At that time, many opinions—shaped by linear thinking—predicted that AI would never advance to the point where it could be truly useful. The "recent rise of AI" is not a sudden breakthrough but rather the next stage in its exponential growth. And remember, with each new step, the progress doubles again.

To grasp the power of exponential thinking, consider the work of one of the most renowned science-fiction writers, Stanisław Lem. Known for his science-informed "hard science fiction," Lem wrote "Golem XIV" in 1981. In the book, the Golem is an AI that surpasses human capabilities and views humans as too simple to engage with. The AI gives a series of lectures, hoping to intellectually awaken humanity. The fact that Lem wrote this over 40 years ago, in communist Poland, highlights the remarkable foresight that exponential thinking can offer. The Golem realized that our civilisation is not capable of understanding him, and ceased speaking to humans in 2047. In 1980's it seemed absurd, but nowadays the book feels strikingly relevant.

The power of digital information drives progress not only in computing but also in other fields. The development of artificial intelligence, robotics, nanotechnology, and many other fields will accelerate, resulting in new challenges and opportunities across all areas of science. What will your research field look like in the future? Try to answer this question while keeping in mind that predicting the future requires abandoning linear thinking.

Science is also becoming increasingly interdisciplinary. The conventional division into disciplines blurs every day. You can see examples of this happening in certain fields: econophysics, bioinformatics, biomimetics—the list goes on. Traditional divisions are a barrier resulting from our need to organise everything. That's psychology. If we can fit something neatly into a box, we feel that we understand it better. However, a negative effect of the division into disciplines is the "grouping" of scientists into specific teams. Chemists talk to chemists and work in the field of chemistry. By breaking down the barriers, you open your mind, learn new things, and get priceless inspirations. You travel in extensive intellectual space not restricted by any human-created border controls.

Where Should Ronaldo Run?

In light of these changes, can you imagine what your research field will look like by the time you complete your Ph.D.? And what about several years later—will it even resemble its current form? When planning the long-term development of your career, it's essential to consider both the latest trends and those expected in the future. Doing so will enable you to stay up to date and continue conducting groundbreaking research that aligns with the future state-of-the-art.

Where does Ronaldo run to pick up a pass? Does he run towards where the ball is at the moment, at the feet of his teammate? Or does he look for a place where the ball could be in a moment? Be like Ronaldo—look for an open space where you can take the ball and make your shot. Think how the pitch will look like in the future.

To finish off, here's a quote from the famous hockey player Wayne Gretzky who once said: *"I skate to where the puck is going to be, not where it has been."*

I love Wayne and, frankly, I wanted to start this section with his quote. But one of the Ph.D. students took a look at the outline and asked me "Who's Wayne Gretzky?" And here I found myself writing a section about being innovative and thriving as a young scientist and emphasising this with a reference to my own youth of the '80 s and '90 s. That's a pretty good example of how easy it is to ignore the fact the world is changing and stick to old habits.

Takeaway points:

- **Technology grows exponentially—after 24 months, it will be twice as advanced as it is today.**
- **Disciplines create artificial barriers—breaking them is a good opportunity for interesting discoveries.**
- **Think about your discipline once you finish your Ph.D. and beyond—how will it look like?**

Chapter 3
Craft Impactful Research Papers

> Research papers are more than words—they're intellectual blueprints and a reflection of your legacy.

3.1 Why Do You Want to Publish?

Share Your Knowledge

Doing impactful research is a significant uphill battle. But, unfortunately, it's not the end of the war. You still have to get it published. Let me be honest, your first research paper will be a huge undertaking. Scientific publications should present (or refer to) new results, be publicly available, clear and concise in the communication. Research papers aren't there to entertain or present sophisticated forms of writing, but to share knowledge—benefiting both the author and the reader. Some scientists say that the study isn't complete until the results are published. A bit like artist painters and singers, their work cannot be discovered if they do not release it through galleries or concerts. Just as a painter seeks to display their work in renowned galleries, you too want your research published in the best journals. Without this, all the hard work you put in the lab may go unnoticed.

So, how can you ensure your paper gets published successfully?

History of Publishing

Before we answer this question, let's go back in time a little bit. The history of scholarly publishing goes back as far as 17th-century France and England, where the first scientific journals were established. Initially, only the editors of the journals were responsible for the selection and evaluation of the content. Over time, the so-called peer-review system was developed, in which works are reviewed by "peers" from a

---

Supplementary Information The online version contains supplementary material available at https://doi.org/10.1007/978-3-031-90102-7_3.

given discipline. Another significant change was the diversification of income streams for academic journals, which contributed to their broader accessibility. In the past, academic journals were often not generating much profit, and many publishers faced financial challenges. The Information Revolution transformed this landscape, leading to the proliferation of journals, the growth of for-profit academic publishing, and the rise of the open-access movement—a model where papers are freely accessible to anyone with an internet connection, but the costs of publication are typically borne by the authors.

There's no need to be modest about your scientific achievements. Scientific publishing is not showing off, but the act of sharing your work. You can reach a wide audience and contribute to scientific progress in your field, open the door to constructive discussion, or inspire further researchers to follow in your footsteps. By increasing your scientific output, you not only gain greater recognition and reputation in the community, but also increase your chances of receiving a grant. Each published paper broadens your experience, skills, and professional contacts.

However, it's important to keep in mind that quality is just as crucial as quantity. By publishing a weak or unprepared paper, you are risking harming your reputation. Someone might reach only to this one unprepared paper and assume that the rest of your work is similar. So you need to find the right balance. Don't do salami publishing, but also don't wait until your publication reaches Nobel Prize level.

Types of Research Papers

Once you are ready to share your findings, you can choose from several publication formats. This could include a research paper, but also a book, an instructional manual, a handbook, conference materials, a dissertation, or a statistical report. It's important to distinguish between these different types of publications, as they each serve distinct purposes. Here are examples of publication types you may consider during your Ph.D.:

- **Full research papers**—They present original findings within the context of existing work in the field. These are typically characterised by a defined structure, most commonly the IMRaD format, which will be discussed in Sect. 3.2.
- **Review papers**—They provide a broad overview of the existing literature on a specific topic. Reviews do not present new findings but require an in-depth understanding of the field. When entering a new research area, finding a relevant review paper is often a good starting point.
- **Papers in a special issue**—Special Issues are usually invited by the editorial office or submitted in response to an open call. These may include papers presented at a specific conference. Another scenario occurs when a journal invites an expert to curate a collection of papers on a selected topic.
- **Short reports or letters**—All kinds of opinion-forming papers where authors express their views on a given issue, supporting their stance with relevant results. Sometimes, these papers may present new studies conducted to support the opinion presented, but it is not required in that form.

- **Communications**—Brief papers that focus on new research or developments in the field. These are typically published more quickly than regular papers, allowing authors to promptly inform the scientific community about their findings.
- **Technical notes**—Concise papers that typically describe new experimental procedures or testing methods. The purpose of a technical note is to provide enough detail so other researchers can replicate the method.
- **Self-publishing**—Papers you publish on your own, bypassing the peer review process. Institutions may publish reports or white papers on their websites. Individual researchers can write online articles for platforms like blogs, LinkedIn, or other websites.

With such a wide range of options for presenting your research, you're sure to find a format that fits your needs and style.

Takeaway points:

- Don't be selfish—share your findings instead of keeping them for yourself.
- **Each good publication improves your credibility, but weak publications can harm it.**
- **Use the whole spectrum by picking the right publication type for your message.**

3.2 What's Your Story?

I'm a Serious Researcher!

Former editor of Science journal, Katrina Kelner, once said: *"The most successful papers are those that present innovative research. But the best papers also present their story in a clear and logical way."*

You might say: "This is science! Is it not enough to do my job? I generate solid results, prepare detailed figures, and analyse them thoroughly. After all, research is a serious thing, and I am not a fiction book writer!" Well, if you want to write something outstanding, you do have to make sure it has an interesting plot to draw the audience in. It's not an obvious approach for scientists, but the benefits of using stories have been proven in many fields. The best research papers take the reader on the journey with the author.

People who have had the greatest impact on our world—be they business leaders, politicians, athletes, or celebrities—usually can capture the masses through telling interesting stories. Stories are one of the most efficient ways of getting your message delivered to the audience. They have been in our lives for millions of years. As a species, we think with stories—they stimulate our emotions and inspire us. So yes, you do serious research, but if you develop your storytelling skills, you will do your job even better.

Story Flow

Your paper begins with the Introduction section. A strong introduction sets the stage with a topic or issue that resonates with the reader. It is here that you define the research question, refer to the relevant literature that frames your study, and explain the theoretical basis of your work. The Introduction should also hint at the conclusion by briefly mentioning the results. In Sect. 3.5, I will "introduce" you to the Introduction.

From there, you move into the Methods and Results sections, where you present your empirical evidence and justify your methodology. These sections form the backbone of your paper, detailing the processes and findings that support your narrative. Methods describes every important practical detail of your work. Well written Methods section should allow the reader to replicate your study. Then the Results section describes what you measured or observed.

As you approach the Discussion section, you broaden the context of your findings by comparing them with existing literature. The climax of the story should occur in the latter part of the Discussion, where you provide a clear and solid answer to the research question posed in the Introduction. This is the moment where the implications of your work become evident to the reader. In Sect. 3.6, I will "discuss" the Discussion with you.

This set of headings forms the most common structure for papers in STEM subjects, known as IMRaD. Yes, you guessed it—the name comes from the first letters of the headings: Introduction, Methods, Results, and Discussion.

Finally, in the Conclusion, you summarise the key takeaways from your study in a concise and accessible way, ensuring the reader remembers the core message of your work.

Story and Core Message: The Process

Planning the plot of the paper should happen at the beginning of the writing process. You should do this once you've reviewed the literature and know what results you have and what conclusions you can draw from them. All you need is a clean sheet of paper and a moment of peace for reflection. It's worth starting by outlining the context of the paper, focusing on:

- **The state of knowledge**—Which scientific domain are you in? What research problem do you analyse? What are the results of other researchers in this area?
- **Your research**—What research method did you use? What results have you obtained? What are the conclusions?

The story of your paper should lead to the most important conclusion: the key takeaway that the reader will remember after finishing the reading it. I refer to this as "The paper's core message". To craft a compelling story, you need to have this main message clearly defined.

You want to ensure that your paper's core message is significant enough to warrant the effort of writing it. Devoting time to a paper with a trivial message is rarely worth

3.2 What's Your Story?

it. To test the relevance and accuracy of your core message, consider asking yourself the following questions:

- What will this paper, and its main message, contribute to my scientific field?
- How does it fit into a wider context? For example, how might it benefit society, industry, or charitable organisations?
- In what ways will it enrich my scientific output and personal research portfolio?

If your answers to these three questions are satisfying and confirm the value of your core message, you can start building paper structure. However, if you're not convinced that your work offers significant value, it's better to invest additional time in crafting the paper's message. Revisit the initial steps of the process, reevaluate your thought process, and refine your ideas until you are confident in the contribution your work makes.

Having the core message defined you know where your story should lead. Now you can proceed to outline the structure of your paper by establishing a sequence of main components:

- **Context**—Summarise the state of knowledge and your research's relevance, derived from the previous step. (Paper section: Introduction)
- **Problem**—Clearly articulate the research question you aim to answer or the hypothesis you intend to test. (Paper section: Final part of the Introduction)
- **Events**—Detail the methodology and results, explaining how the research was conducted and what was discovered. (Paper sections: Methods and Results)
- **Peak**—Highlight the turning point of the narrative, where the meaning of the findings and their connections become apparent. (Paper section: Discussion)
- **Solution**—Clarify the significance of the observed relationships and their implications. (Paper section: Final part of the Discussion)
- **Closing**—Conclude the story by reiterating the most critical observations. (Paper section: Conclusions)

To assist you in defining the story and core message of your paper, we have created the Research Paper Canvas and the Research Paper Core Message Map. Both resources are available for free download from the Springer Nature book site.

Takeaway points:

- **Your paper will stand out if you present your results using a clear and logical story.**
- **Plan your story before you start writing the paper.**
- **A good story leads to one core message presenting clear academic value.**

3.3 Your Title Can Make or Break Your Paper

Why Worry About the Title?

Many years of experience as a journal editor showed me how little authors think about the title of their papers. You don't even have to be an editor to see that. You can just look through the last few editions of a journal of your choice to see what I mean.

So what makes for an "effective title" and why should you worry about it? By my definition, an effective title is the one which:

- Will actively encourage the reader to read the abstract.
- Stands out from the others in a given issue of the journal.
- Is impossible to be ignored.

We live in a world where we must compete for our readers' attention and convince them that our paper is worth their time. A good title is the hook you need to draw them in.

The challenge lies in crafting a title compelling enough to grab the reader's attention and make them want to read the abstract without hesitation. A useful exercise to test the effectiveness of a title is to imagine yourself as a reader skimming through the titles of an entire journal issue. With around forty titles to choose from, the likelihood of a reader selecting your paper is just 2.5% if all the titles are equally appealing. To increase this probability, you need a title that stands out—something that piques curiosity and conveys the value of your work.

Examples of Titles

Step one to writing a great title is identifying the key elements of your research that make it unique and engaging. Let's use an example in the field of materials engineering. Imagine you've developed a super-efficient adhesive tape inspired by the remarkable adhesion of gecko's feet, which allow them to climb vertical surfaces and ceilings effortlessly. The title of your paper could emphasise both the innovation and the bionics inspiration. For instance:

> "Optimisation of stiffness and density of microscopic carbon and glass fibres up to 5 micrometres in diameter inspired by the spatial organisation and hierarchical structure of the wood gecko's foot."

This title certainly contains a lot of information, but it doesn't meet my definition of an effective title. First of all, it's far too long and unnecessarily contains a lot of technical information that would be better presented in the abstract.

Let's try again:

> "From gecko to Spider-Man: the influence of the hierarchical structure of the gecko's foot on adhesion to rough surfaces."

This title combines scientific relevance with an element of intrigue, making it more likely to catch a reader's attention. Mentioning Spider-Man adds a layer of relatability and curiosity, especially for non-specialist readers or those casually browsing.

While it's true that such a title might seem less formal, it aligns with the primary goal of attracting readers in an increasingly competitive publishing landscape. Scientific journals generally allow for creative freedom in titles as long as the content of the paper maintains rigor and quality. By striking this balance, you can make your research more accessible and engaging without compromising its scientific integrity.

However, for those who want to get rid of Spider-Man, I have another idea:

"Producing a material inspired by the adhesion of the gecko's foot for use on everyday surfaces."

Or something even shorter:

"Creating gecko-like adhesives for 'real world' surfaces."

This just so happens to be the title of an actual paper published in Advanced Materials journal. It's short, sweet, and does the job of letting the reader know exactly what they need to without be overwhelming them.

Conscious Decision

The link between the length of the title and the paper popularity (expressed in the number of citations it receives) has been studied by scientists from the University of Warwick. And, rather aptly, their findings were published in the paper "The advantage of short paper titles". Based on more than 100,000 analysed papers from different journals, the authors found the following relation: the fewer words in the title, the more often such paper was cited by other authors.

To sum up, I encourage you to experiment with your titles and try to choose one that no reader could pass by. The worst-case scenario is the editor will politely ask you to change your title. It's not as if they're going to reject it. The opposite worst-case scenario is you choose a poor title, lose readers and, consequently, potential impact. So why not try something different?

Takeaway points:

- **You're competing for the reader's attention, use your titles to win them over.**
- **Craft your paper's title carefully instead of using a long, dull one with many keywords.**
- **Papers with short titles tend to be cited more often than papers with long titles.**

3.4 How to Write a Captivating Abstract in Ten Minutes

What Makes a Good Abstract?

Let's start by looking at what distinguishes a good abstract from a bad one. Remember, abstract is what will make the reader want to consume or skip the whole paper. A great title isn't the end of the battle, after all. A good abstract draws the reader into the story of the paper. It should explain why the research was conducted, why the results are important, and present the main conclusions from the paper. It arouses the reader's interest, at the same time leaving them hungry for more.

What happens if you don't write a good abstract? Sure, the world probably won't end, but also your paper won't get any eyes on it. And since you've already done all the hard work—research and writing the paper—you want to reach as many readers as possible. It's a path to impact, citations, new contacts, and invitations to conferences.

While some journals provide a specific structure for abstracts, these guidelines alone won't ensure an abstract that captivates your readers. As the author, the responsibility of crafting an engaging and effective abstract lies squarely on your shoulders. To help you succeed, I recommend a framework developed by Professor Patrick Dunleavy, which has proven to be highly practical and insightful.

Five-Ingredient Recipe

The framework consists of writing several sentences for each of the following points:

- **Works of other authors from the thematic area of your paper**—Place your work in the broader context of the research area (max 50 words).
- **What's special about your approach to the problem?**—Explain how your work differs from the existing studies (max 40 words).
- **Your research method**—Describe the methods you used (from 40 to 120 words, depending on how much detail is needed to reliably describe the methods).
- **Main conclusions from the work**—Describe the consequences of your results. Avoid vague or general statements, and try to be specific and concise. The length of this section will depend on the word limit for the entire abstract and the number of words used to describe the research methods.
- **Added value of your work**—Explain why your work is particularly original and important in your narrow research field. If the reader remembers only one thing from your paper, it should be this point (max 30 words).

At this point, your abstract is almost ready. There are just two more things to check. First, consistency: Does the text of the abstract align with the title? Do the title and the first three sentences make the reader want to continue reading? Secondly, you should check its conciseness. Are there any redundant words? Which words or phrases hinder the clarity of the main message?

Finally, one last remark: it's best to write the abstract after writing the entire paper. Only then will you be able to effectively describe each of the five points above.

Takeaway points:

- **The most succesful authors craft their abstracts very carefully.**
- **Like titles, abstracts are crucial for attracting the reader's attention and encouraging them to read your paper.**
- **Use the structured set of five points to write an effective abstract.**

3.5 Writing Your Introduction in Seven Easy Steps

Pragmatic Approach

In my opinion, after the Discussion section, the Introduction is the most challenging part of a research paper to write. The Introduction section aims to describe the research topic and demonstrate its significance. You start presenting state of the art from a general big picture. Then you become more specific and detailed to describe a particular research area. After describing 3–5 key papers related to your work, it highlights the existing gap that needs to be addressed. Finally, many researchers close up the section by specifically stating the aim of the current study, which is supposed to close that gap.

You need to showcase a deep understanding of the subject and convey it in a concise, engaging way. As a pragmatist, I prefer breaking down complex problems into manageable parts and addressing them systematically. This approach transforms the daunting large task into a series of well-defined steps that can be planned and executed over time. It's a method I use for tackling the Introduction section, and I've found it particularly effective when working with Ph.D. students.

Solid Literature Review

The prerequisite for the creation of the Introduction is an extensive review of the literature. However, before you start reading or writing anything, it's a good idea to set a deadline for when the literature review and the Introduction should be ready. Literature review is one of those things that would never be finished. That's why it's important to treat your deadline seriously and keep yourself committed to it. That's why you should involve a third party, such as a co-author or your supervisor to commit to the deadline. This way, you aren't beholden to yourself—someone else is relying on you. It's a subconscious way to push yourself to achieve your goal and stop you from putting it off forever.

Something else that can help you, is to block a few two- to four-hour sessions in your calendar over the next few days. Half an hour is definitely not enough to get into the subject and do valuable work, while spending the whole day on one task might be impossible for a busy Ph.D. student.

Seven Steps

Remember how I mentioned we were going to break the introduction down into parts? Well, here they are—seven easy steps that will make writing that intro a little less daunting.

1. **Collecting your literature**—Here's where you need to identify references relevant to your paper. In the Introduction, it's important to start broadly and gradually narrow the focus, so you'll need to gather papers that provide both general context and specific background for your research. Depending on your preferred workflow, you can collect references in printed form or use a reference management tool. At this stage, aim to identify around 100 papers, though the exact number

will vary depending on your field. The assumption is that about one-third of these will survive the selection process later on and be referenced in your paper.
2. **Grouping your papers**—The next step is to conduct a quick overview of all the papers—a 'shallow reading.' The aim is to group the references by topic. Some papers might provide an introduction to a research field, others could offer examples of specific research methods, and a few may present results similar to your own. This step is crucial as it helps you determine the structure of your entire Introduction section. Remember to follow the principle of moving "from general to specific." Start by outlining the broader area of research, then narrow down to define the problem. It's most effective to complete this task in one focused session.
3. **Mapping out your headlines**—Now you can create a document for your Introduction, but for now, you'll only include the headings that align with the reference groups you established in the previous step. The exact wording of these headings isn't important, as they'll be removed later. Instead, focus on creating concise keywords that will help you outline a coherent structure. Ideally, aim for four to eight headings. And, to reiterate, ensure the progression flows from general to specific!
4. **Selecting key papers**—After dividing the Introduction into sections, you can focus on each one separately, with the aim of turning each section into a single paragraph. It's best to tackle this task incrementally, working on each section during the focused blocks of time you've set aside. Leave the overall Introduction document for now and, instead, conduct a thorough, detailed reading ("deep read") of the selected papers for each section. The goal is to identify the most relevant papers that will feature in your Introduction and be added to your reference list. As you read, highlight key text, add annotations, or take notes to streamline the writing process later.
5. **Creating your paragraph structure**—Now, return to your Introduction document and, using the notes you've taken from the references, create a list of key points to cover in each section. Each section should have its own dedicated list of bullet points. These lists will serve as the foundation for your paragraphs, so ensure their wording and order create a coherent and logical flow. Keep in mind that each paragraph should follow a clear structure: an introduction to the topic, an expansion of the idea, and a concluding sentence that ties it together.
6. **Converting points into sentences**—Now, you can begin the actual writing process. Start by transforming the individual points into complete sentences, incorporating the relevant references as you go. It's helpful to tackle this task section by section, focusing on crafting one paragraph at a time before moving on to the next.
7. **Checking the introduction**—The final step is to remove the header titles and refine the overall flow of the Introduction. At this stage, your focus shifts to the style and readability of the text rather than the content. Ensure that the entire section flows smoothly and is easy to follow.

Writing a research paper is a long and meticulous process. Rushing through it without proper preparation won't yield quality results or produce a paper you can be proud of. A well-crafted Introduction cannot be written in a single sitting—it's far too important for that. If your title and abstract have successfully captured the reader's interest, the last thing you want is to lose them in the Introduction. With careful planning and by breaking the work into manageable steps, reviewing the literature and writing the Introduction can become a much more rewarding experience. Who knows, you might even find it enjoyable!

Takeaway points:

- **Don't start writing your Introduction before you have a solid understanding of the background literature.**
- **Forget about writing it in one go; this process requires time.**
- **Breaking the process into seven easy steps will allow you to plan and execute it efficiently.**

3.6 Let's Have a Meaningful Discussion

This Is It: The Most Important Part of the Paper

The aim of the Discussion section is to interpret and describe the significance of your results in the light of the current state of knowledge. The Discussion should connect back to the Introduction by addressing the research questions or hypotheses and incorporating relevant references. This is where you explain how your research has contributed to the field and enhanced the understanding of the research problem.

Try searching "IMRaD" images online, you'll find visual representations of this structure in the form of an hourglass. The Introduction and Discussion sections create this shape together. While the Introduction moves from general to specific (hence, it's represented with a converging shape), the Discussion works in the opposite direction, broadening the context of your findings and exploring their implications for the state of the art (which is why it's represented with a diverging shape).

Description Versus Interpretation

Discussion is often seen as the crucial component of the research work, because it demonstrates the author's ability to think critically. It also showcases an original contribution based on the research results and formulates a deeper understanding of the problem.

It is where the author must present the significance of their research and explain how it contributes to addressing existing gaps in the field. If necessary, the Discussion should also highlight how the research results have revealed new gaps in the literature, that were previously unnoticed or inadequately explored.

The Discussion section goes beyond merely presenting objective facts. It offers a space for creative, evidence-based interpretation of your findings. This is where you can give your results context and explore their broader implications, offering a deeper understanding of the research problem.

The following list of phrases illustrates two approaches to writing. The first approach is descriptive, often found in the Results section, while the second is more analytical, delving deeper into interpreting the results, which is characteristic of the Discussion section.

Description	Interpretation
States what happened	Identifies the significance
States what something is like	Evaluates strengths and weaknesses
Gives the story so far	Weighs one piece of information against another
States the order in which things happened	Makes reasoned judgments
Says how to do something	Argues a case according to evidence
Explains what a theory says	Shows why something is relevant or suitable
Explains how something works	Indicates why something will work best
Notes the method used	Indicates whether something is appropriate
Says when something occurred	Identifies why the timing is important
States options	Gives reasons for selecting each option
Lists details	Evaluates the relative significance of details
States links between items	Shows links between pieces of information
Gives information	Draws conclusions

Structure and Problems to Avoid

The main components of a Discussion section in a research paper are:

- **Summary of key findings**—Highlight the most important results of your study.
- **Interpretation of results**—Explain what your findings mean and their significance in the context of your research question.
- **Comparison with previous work**—Relate your results to existing literature, showing how they fit in or differ from prior studies.
- **Implications and significance**—Discuss why your results matter and their potential impact on the field.
- **Limitations**—Address any weaknesses or shortcomings of your study and how they may have affected the results.
- **Recommendations for future research**—Suggest avenues for further studies or analyses based on your findings.

One of the most common mistakes when writing the Discussion section is presenting a superficial interpretation of the results, essentially just reiterating the content from the Results section. While it's necessary to reference the results, the focus should shift towards interpreting these findings, not merely describing them again. In the Results section, you present the "WHAT HAPPENED"; in the Discussion, you should address the "WHAT IT MEANS". Sometimes I see that authors combine the two sections into "Results and Discussion". Although it can be a good approach, I've seen many authors use it to disguise a lack of deep discussion in their papers. Sometimes it was done on purpose, but in most cases authors were not aware of the necessity of having deep discussion beyond the context of the current paper.

By following the steps outlined here, you'll be able to bring greater meaning to your results. This is what really makes an outstanding paper! This is also the moment when you can fully express everything you've learned and discovered in your research area. After all the time spent analysing and reading, it's now your chance to contribute to the ongoing conversation. Let your passion shine through and demonstrate that you have valuable insights to share.

Takeaway points:

- **Be under no illusion, the Discussion section is the most difficult to write.**
- **Make sure you interpret your results and not only describe them.**
- **A strong Discussion can be the difference between an average paper and an outstanding one.**

3.7 Nine Ways to Break Your White Page Syndrome

You Are Not Alone

Do you struggle with focusing on writing? Do you experience "white page" syndrome, where you sit staring at a blank page, hoping for the perfect first sentence to come to you? Does the thought of writing an entire paper feel overwhelming? Welcome to the club. You're in great company. Despite their fame, even the most renowned authors face challenges with writing. James Joyce once said in a letter that *"writing in English is one of the most ingenious tortures for sins committed in the past"*. Similarly, John Dos Passos joked in a New York Times interview that if writers were assigned to a special level of hell after death, it would be one where they must endlessly contemplate their own work.

Many authors have created outstanding works, but that doesn't mean their journey was smooth. Writing research papers requires hard work and perseverance, so don't feel discouraged when you freeze at the thought of this monumental task. If you wish it was easier, allow me to suggest some ways to improve the efficiency of this process and make it more manageable.

Let's Fill the Page!

Ready? Steady? Go! You can start tackling the blank page syndrome now by using those proven strategies.

Look at the big picture—Start by sketching the story you want to tell. Whether it's mapping out the overall structure on a blank sheet or arranging a series of research results in chronological order, the key is to begin with the big picture.

Clear divisions—Once you've outlined the shape of your paper, it's time to focus on the smaller tasks. Large projects can be overwhelming, so break them into manageable chunks. Edit individual parts—each one we've discussed so far—in separate files. Smaller page counts make each task feel less daunting.

Map out your paragraphs—Think of each paragraph as a brick in your writing. Every paragraph should focus on one idea, with a clear structure: an introduction, developing details, and summary. A paragraph that's too short (just two sentences) is incomplete, while one that stretches over a whole page is overwhelming.

Bite-sized goals—It's better to write something imperfect than nothing at all. It's easier to improve poor writing than to start from scratch. Set small daily goals—two paragraphs, for instance. If they're bad, they're bad. Remember, no first draft is perfect!

Treat yourself—Once you hit your daily goal, reward yourself! Treats can vary from your favourite snack, enjoying a social media break, a walk, or calling your loved one. Choose what makes you happiest. Motivation is key, and you deserve to celebrate your progress.

Natural rhythm—Work with your natural rhythm. For most, morning is the most productive time, but if you're a night owl, follow that schedule instead. Avoid the trap of writing all day—it will only burn you out. Focus for two hours when you're most productive, rather than forcing yourself to stretch past your attention span.

Avoid distractions—Distractions are your enemy. You know what pulls your attention away—emails, social media, etc. Close your inbox, hide your phone, or even disconnect from the internet entirely. Put your devices in airplane mode and enjoy a few distraction-free hours to concentrate fully.

The Pomodoro Technique—It's not pasta, but it's a great productivity tool. The Pomodoro Technique involves focused work for 25 min, followed by a 5-min break. During the break, stretch, get a drink, or take a quick walk. This rhythm aligns with your concentration abilities and can significantly boost productivity. More on Pomodoro Technique in Sect. 7.7.

Sounds—Achieve deep focus by listening to sounds that enhance your concentration. Classical music, white noise, brown noise, or specific "concentration music" playlists on YouTube may work best for you. Experiment and find what helps you zone in.

Create Your Own Ritual

These tips stem from many years of personal experience. I've often wondered why some authors can churn out research papers one after another, while others struggle, spending weeks on a single piece and draining themselves in the process. The biggest obstacle in completing what we start is often ourselves—our doubts, our weaknesses, and our failure to plan effectively. But don't let these hurdles stop you. Decide today that you are an effective author, and that you will produce great work.

Not all techniques will work for you. Treat yourself as an experiment and try different methods one after another to see which yields the best results. By doing so, you'll discover your own unique writing ritual. It's up to you to create the ideal environment that enables you to write at your best.

Takeaway points:

- **It's not just you. Writing is tough for everyone.**
- **Mastering the art of writing is essential for making meaningful progress in your Ph.D. (and life).**
- **See what works best for you from the above list and tailor it into your own writing practice.**

3.8 Plagiarism in Scholarly Publishing

Recycling Is Not Always a Good Thing

Both scientific publishers and journal editors are highly sensitive to plagiarism. This concern extends beyond authors copying the work of others or using large language models in ways that violate publishers' guidelines. Plagiarism is more complex than that. In fact, you can plagiarise yourself, something that's often referred to as "text recycling".

Let's look at an example. An author is finally sitting down to write a paper on their latest research. They're laying out their sections and come to the Methods. Since this research employed the same experiments as their previous paper, the author decides to copy and paste that section into the new paper. What's the harm in that? After all—it's their own text. That simple assumption is going to land that researcher in hot water.

This is a classic case of auto-plagiarism, or text recycling. To understand why this is problematic, it's important to know how copyright works in traditional model of scholarly publishing. When an author submits a paper and it is accepted for publication, they typically transfer the copyright of their work to the publisher. This means that while the author initially wrote the text, the copyright now belongs to the publisher. In return, the publisher takes on the costs and responsibilities of preparing the paper for publication.

However, publishers require exclusive ownership of the material they publish. If the author reuses sections from a previously published paper, the new submission violates copyright agreement of the previous one. This creates a legal and ethical dilemma for both the author and the publisher.

The same principles apply to the reuse of images and graphics, though detecting image plagiarism can be more difficult. However, the consequences are just as serious if it is discovered. To avoid these issues, the rule is simple: when in doubt, write it out! Always aim to create original content for each submission, even if you are working with familiar methods or concepts. You are also allowed to reuse the same data in a different graphical form. What is protected is the graphical representation of the data. The raw data itself cannot be copyrighted and can be reused to create new figures.

The Role of the Trusty Editor

Now let's have a look at a traditional plagiarism—copying others' work. Software such as Turnitin or CrossCheck can easily detect such instances. Authors should assume that every publisher employs these tools to identify plagiarism. Anti-plagiarism software acts like a search engine, comparing the content of a paper with resources available online, such as other research papers, websites, and e-books. It identifies identical sequences of text and calculates a compliance percentage, indicating how much of the text overlaps with existing sources.

This compliance percentage serves as a guide for editors, who must interpret the results before making decisions about the paper. Most submissions have a compliance percentage of 5–10%, which is considered normal. This is because scientific writing often includes universal phrases, commonly used terminology, or shared formulations. Additionally, references and citations can sometimes appear as sources of overlap, which is a misleading artifact of the software.

Editors typically become concerned when the compliance percentage exceeds a certain threshold, such as 15%, though this varies by discipline. When in doubt, the editor will review a detailed report delivered by the anti-plagiarism software. It allows to identify problematic sections and pinpoint used sources. Editor's primary role is to determine whether the overlap reflects intentional plagiarism or unintentional misuse of quotations or citations. They generally disregard isolated clusters of three or four words; the concern arises when entire paragraphs closely resemble previously published texts.

This nuanced approach helps editors balance the detection of potential misconduct with the recognition of acceptable and unavoidable similarities in academic writing.

Use of AI

This is relatively new and hot topic in academic publishing. The landscape is changing rapidly, so the best idea is to consult the latest publisher's guidance. In general, it is allowed to use "AI-assisted copy editing", which refers to the use of artificial intelligence tools to enhance the clarity, readability, and correctness of human-generated texts. Usually, such assistance does not require formal declaration, as it is viewed as a tool for refinement rather than substantive content generation or creative contribution.

AI-assisted copy editing involves a range of tasks such as: correcting grammar, spelling, and punctuation errors, ensuring consistency in tone and style, and adjusting wording and formatting to improve readability. For example, an AI tool might rephrase awkward sentences, reorganise sentence structures to enhance flow, or standardise terminology usage throughout a paper. However, it is crucial to distinguish AI-assisted copy editing from more substantial forms of editorial involvement, such as generative writing, where AI creates original content autonomously. While copy editing focuses on refining and polishing existing text, generative writing involves AI producing content from scratch, a step that goes beyond the scope of simple editing.

Researchers can benefit significantly from AI-assisted copy editing by improving the clarity of their work without compromising its originality or authorship. This is especially a case when English is not their first language. AI becomes a partner, helping you to produce polished, error-free, and coherent texts while leaving the core intellectual contribution firmly in your hands.

Takeaway points:

- **This is non-negotiable: do not plagiarise another author's work.**
- **Remember, re-using your own text and figures published in another paper is also considered as plagiarism.**
- **You can use the AI agents responsibly.**

3.9 Are You a LaTeX Geek?

Why LaTeX?

The vast majority of researchers write papers in Microsoft Word. However, this isn't the only tool available. There's no shortage of alternative word processors that work just as well, if not better, than Word: Scrivener, Open Office, Google Documents, DropBox Paper, or the subject of this section—LaTeX.

LaTeX is an open-source platform for writing documents, especially popular among scientists and researchers in STEM subjects. Many use it for a variety of tasks, such as writing research papers, books, M.Sc./Ph.D. theses, presentations, posters, and more. LaTeX has garnered a strong enthusiasts, but it also has its detractors. Let's explore some of the key pros and cons of LaTeX.

Pros and Cons

What makes LaTeX so popular with scientists? Especially over more conventional word processors? Let's take a look at the main advantages.

- **Focus on content**—When writing with LaTeX, you don't have to worry about font size, margins, tab size, or other formatting details. The environment is designed so that the visual layer is separated from the text itself. This allows you to focus on the content of your paper, leaving the formatting for later.
- **Efficiency**—LaTeX allows you to automatically reference other elements in your document. With a single command, you can refer to a figure, table, equation, section, or citation. The numbering of these elements is automatic and updates as you create the document.
- **Journal styles**—LaTeX allows you to use external files to quickly modify the layout and format of your paper. For example, by replacing a style file you can alter the visuals of the entire document. The same applies to reference formatting. Many journals provide their style files and bibliography formats, so by simply implementing these files, you can adapt the appearance of your paper in seconds.
- **Mathematical formulas**—LaTeX uses its own language for writing mathematical formulas. This may require some initial learning. However, the process is quick because the syntax is simple and intuitive. You may have already encountered it, as it's used in tools like Matlab, Wikipedia, and even as a plug-in for Microsoft

Word. This approach allows you to write smoothly without interrupting your workflow or losing focus. Once you've mastered it, writing becomes faster, and your formulas will look professional.
- **Presentation**—Not much to say here; LaTeX papers look great! The generated PDF is clear and professional.

Despite its benefits, LaTeX does have its detractors. Here's what many of them don't like about the platform.
- **WYSIWYG**—The main difference between Word and LaTeX is that LaTeX is not a WYSIWYG (What You See Is What You Get) editor. In LaTeX, you don't edit directly in the document, and style changes aren't immediately visible. In contrast, Word provides an intuitive interface where you can see the changes in real-time. Writing in LaTeX involves switching between two windows: one for your raw text and the other for the rendered appearance of the document.
- **Compilation**—The second challenge is the need to compile the text in LaTeX. Similar to programming languages, LaTeX requires you to write the content first, then compile it to see the updated appearance of the paper.
- **Input barrier**—Writing in LaTeX is relatively easy and intuitive, but it does require knowledge of s commands. This can be a barrier to entry for many as it's much easier to just start writing.

LaTeX Editors

You can ease the learning curve by choosing a good editor that intuitively supports text and command structures. There are plenty of options available, including desktop apps: Texmaker, LyX, TeXstudio, Kile, and online editors: Overleaf, Papeeria.

I use Overleaf, which allows me to conveniently share the editing window with the paper preview. It also supports group work, comments, online discussions, and even direct submission to certain journals. Integrated with TikZ, one of the most powerful tools for creating graphics in LaTeX, Overleaf is a highly recommended cloud-based editor for writing, editing, and publishing scientific documents. It's an excellent choice for those just starting their LaTeX journey.

Which group will you fall into? Will you embrace LaTeX's distraction-free interface, or will you find that it overcomplicates the writing process? There's only one way to find out.

Takeaway points:
- **LaTeX is designed to focus on content and still produce beautiful documents.**
- **After overcoming the entry barrier, you'll wonder how you ever lived without LaTeX.**
- **You can start with a user-friendly LaTeX editor.**

Chapter 4
Beyond the Draft: Get Published

> Your research isn't finished until it finds its audience and sparks conversation.

4.1 Understanding Peer-Review

The Rules of the Game

Before a paper is accepted for publication in a journal, it must undergo a quality assessment and compliance check with the journal's standards. This process is called "peer-review". A peer in this scenario is a researcher in a similar field, and a review involves evaluating your work. In the context of scientific journals, peer-review process is tailored to assess the quality, validity, and accuracy of a paper before publication. The reviewers are experts in the subject area of the paper who are actively publishing in the field.

The process of evaluating a paper can vary between journals and depends significantly on the reviewer's working style and preferences. However, there are certain fixed elements that are consistently assessed during the peer-review process:

- Alignment of the paper's subject with the journal's scope and profile.
- The methodological soundness and validity of the research presented.
- Clarity and coherence in the structure and organisation of the text.
- Effective and appropriate use of figures and tables to support the content.
- Proper spelling, grammar, and overall language quality.

The Two Stages

The peer-review process begins when the author submits their paper to a selected journal. The journal editor reviews the manuscript to determine whether it meets the journal's basic requirements assessing the manuscript's alignment with the journal's scope and subject matter. The editor's primary role here is to filter out submissions that are unsuitable for further review, resulting in what is known as a "desk rejection." This is stage one of the peer-review process, and I will talk more about "desk rejection" in Sect. 4.5.

This approach ensures that reviewers are not burdened with unnecessary work, a crucial consideration given that most reviewers are already snowed under requests to assess manuscripts. Prioritising papers with real potential for publication is a strategic and efficient use of their time.

If the editor deems the manuscript valuable to the journal, they select two or three reviewers with expertise in the papers's subject matter and invite them to provide their detailed assessments. This second stage is the cornerstone of the peer-review process, where a scientist takes on the role of a reviewer to critically evaluate the quality of another scientist's work. Reviewers typically remain anonymous to the author, although they often have access to the author's identity and affiliations.

Why Is It Such a Long Process?

The review process can often be lengthy due to the multiple stages and individuals involved. Reviewers are typically given several weeks to complete their evaluations, though in some cases, this can stretch into months despite reminders from the editors. Additionally, editors may face challenges in finding willing reviewers—this might occur during holiday periods or when the paper's topic is particularly specialised. For authors, the waiting period can feel interminable, but it's important to remember the intricate procedures happening behind the scenes.

On average, authors can expect to receive feedback on their manuscript within two to three months. Some journals have streamlined their processes to deliver outcomes in just a few weeks, while others may take up to a year. The efficiency of the review process largely depends on the volume of submissions, the responsiveness of the editor-in-chief, and the speed of the reviewers. Unfortunately, authors have no control over this timeline and must simply wait for the evaluation. Patience is key—so prepare yourself for the possibility of a long wait!

The peer-review system relies on the voluntary efforts of reviewers, which is one of the reasons the process can be lengthy—they are offering their time without compensation. This dynamic has sparked ethical debates within the scientific and publishing community. Critics argue that publishers profit from the unpaid labour of reviewers while charging high fees for access to published papers.

Why do reviewers take on this work? The answer isn't straightforward but may lie in the reciprocal nature of the system. Reviewers comply with these unwritten rules because they, too, rely on peer reviews for their own publications. It's a system built on mutual exchange: a "you scratch my back, I'll scratch yours" approach that sustains the cycle.

No one has yet proposed a better alternative to the peer-review system for assessing the quality of scientific papers. As Winston Churchill said, *"Democracy is the worst form of government—except for all the others that have been tried"*. Similarly, peer-review may not be perfect, but it remains the most reliable method we have.

Receiving an invitation to review a paper also serves as an appreciation of the reviewer's scientific competence and reinforces their status as a respected expert in their field. It's not just a duty; it's a recognition of their authority and expertise within the academic community.

From the author's perspective, the peer-review system is a bit like a game with established rules that must be learned and followed. Understanding these rules not only makes navigating the process more manageable but also increases the chances of achieving a favourable outcome—successfully publishing your paper.

Takeaway points:

- **Peer-review isn't ideal, but it's the best process we currently have.**
- **Remember about the two stages of peer-review: you need to impress the editor first before your paper even gets send out to reviewers.**
- **Peer-review is a game - understand the rules and be an effective player.**

4.2 The Best Journals for Your Research

Measuring the Impact

When selecting a journal for your research, it's important to consider journal metrics, or at least be aware of them. Evaluating the quality of scientific work is essential, as individual researchers and entire fields compete for limited number of publication spots. Citations serve as a proxy for measuring research quality, and they play a key role in assessing a scientist's reputation (e.g., h-index) as well as the impact of scholarly journals (e.g. Impact Factor, SNIP).

However, there are signs that the citation based system leeds to negative effects such as: the proliferation of mediocre papers, self-citation cartels, and inflated reference lists. This highlights the need for new ways to measure success in science. Research has shown that an overemphasis on metrics tends to focus our attention on factors that are easy to measure, which may not always be the most relevant. This phenomenon is aptly captured by Goodhart's Law, which states, *"When a measure becomes a target, it ceases to be a good measure"*. Current academic productivity measures often prioritise a short-term publication cycle over focusing on long-term research capacity. This creates a risk of a closed loop, where new generations of metrics encourage more ways to game the system. Therefore, I suggest that responsible authors should focus on both sides of the medal: conducting high-quality research but also on making their work easily discoverable and engaging to read, thereby improving its impact.

Where Not to Publish

A good starting point is to understand what journals should be avoided! Have you heard of predatory journals? These are journals that claim to be peer-reviewed but, in reality, publish all submitted papers as long as the author pays a publication fee. The problem is that publishing in such a journal offers no real academic recognition. Leading scientific databases like Scopus and Web of Science do not index these journals, meaning they may be absolutely unknown in the academic community.

Therefore, it is crucial to learn how to identify predatory journals accurately. Here are some red flags to look out for:

- **Contact details**—Does the journal provide verifiable contact information? Can you reach the editorial office, publishing house, or editor-in-chief? Do committee members have email addresses associated with their scientific institutions? Predatory journals often provide fictitious or non-functional contact details.
- **Aims and scope**—Is the journal's scope too broad, covering unrelated scientific fields? Does the editorial board align with the journal's focus? Predatory journals often aim for an overly broad scope to attract as many authors as possible. Since they don't require a rigorous peer-review process, they also don't need qualified reviewers.
- **Scientific committee**—Can you verify the members of the scientific committee through their academic institutions? If possible, reach out to one of the committee members to ask about their experience with the journal. Predatory journals may list committee members who aren't actually involved, or who may have only been contacted once.
- **Peer-review**—Does the journal provide clear details about its peer-review process on its website? Predatory journals typically don't explain the process because they don't conduct proper peer-reviews; they only require authors to pay a fee.
- **Indexing**—Is the journal indexed by reputable databases such as Web of Science or Scopus? Predatory journals often falsely claim indexation in Google Scholar (which is a search engine, not a database) or feature questionable credentials like presence in the Index Copernicus database.
- **Retraction policy**—Does the journal outline the steps it takes for papers that may not meet ethical standards, such as those suspected of plagiarism? Predatory journals usually lack clear retraction policies because they don't enforce quality control.
- **Simplified review**—Does the journal require authors to submit reviews along with their papers? This is an unethical and unacceptable practice in legitimate scientific journals.

All the points above, except for the last one, should be considered general guidelines. A journal may still be of decent quality even if one of the red flags is present. However, if you notice 2–3 red flags on a journal's webpage, you should approach with caution. To be sure, you can check the journal's metrics using verified sources like Scimago, Eigenfactor (free access), or Scopus and Web of Science (paid access, typically available through your library). Additionally, I recommend reading "Who's Afraid of Peer Review?", an article by Science correspondent John Bohannon, which investigates the peer-review practices of fee-charging open-access journals.

Journal Finders

Since you are familiar with journal metrics and know how to avoid publishing in the wrong journals, let's discuss how to choose the right one for your research. Typically, your supervisor and/or co-authors will suggest a journal to submit your work to. However, it's beneficial to understand the broader topic of journal selection and explore different options. This is where journal finders come into play.

Most major publishers, such as Elsevier, Springer, and Wiley, offer journal finder tools. These tools ask you to input key information about your paper, such as the title, abstract, and keywords. The algorithm then compares the terms you've provided with the keywords assigned to different journals, generating a list of journals that are thematically relevant to your work. The search results usually include additional details about each journal, helping you make a well-informed decision.

An undeniable disadvantage of journal finders is that they are often limited to journals from a specific publisher. To broaden your search, it's worth using journal databases. Here's a reminder of the list: Scimago, Eigenfactor, Scopus, and Web of Science. Scimago is likely the easiest to use. It is a publicly accessible portal that provides comprehensive scientific indicators for journals and countries, derived from data in the Scopus database. By using Scimago, you can identify journals within your research area and access a broad range of options across various publishers. This approach will help you make a more diverse and informed selection for your paper submission.

In the search for a suitable journal, it's beneficial to use multiple search engines. Each search tool relies on a different algorithm, which can yield varying results. After analysing all your options, you can create a shortlist of journals with different thematic focuses and credibility. The next crucial step is to visit the websites of these journals. The journal's website is the most reliable and up-to-date source of information about its scope, submission guidelines, peer-review process, and more. While it may require some effort, thoroughly examining each journal's website will ultimately ensure you make an informed and strategic choice for your paper. Last, but not least, you can look through the last couple of issues of the selected journal and imagine if your paper will look good among them.

Takeaway points:

- **Citation-based metrics are flawed, but it's still the most common way for measuring academic impact.**
- **Now you know how to recognise predatory journals, avoid them.**
- **There is handful of tools available to help you find new journals and check their quality.**

4.3 Who Are Your Co-authors?

Quality Versus Quantity

Authorship can be a sensitive and emotionally charged topic. Should there be fewer authors? Who deserves co-authorship? How can conflicts over authorship be avoided? These questions often appear somewhere on the way for the final publication. However, by following some best practices and being mindful of common pitfalls, you can navigate the process more smoothly and ensure clarity and fairness. Here's some guidance to help.

I've encountered many researchers who believe that having fewer co-authors is always better. However, I don't agree with this view. Instead, I suggest focusing on the quality of the publication and how to reach the right audience. The number of authors often depends on the scientific discipline. In the humanities and social sciences, papers typically have fewer authors, whereas in more technical disciplines, more than 10 authors are not unusual. In the medical sciences and physics we observe the phenomenon of "mega-authorship", like papers from CERN with over 5,000 authors.

An interesting correlation was observed in the paper "Impact Factors: Use and Abuse", where researchers compared the average impact factor and the average number of authors within different scientific disciplines. They found a corelation between these two variables. However, this doesn't directly imply that publications with more authors are inherently better. The situation is much more complex. For instance, each scientist has their own network of collaborators, and with more authors, the publication has the potential to reach a larger audience. Additionally, each new author provides another set of eyes to review the results and refine the publication. In this sense, it could be argued that the more authors there are, the better.

Recognising Diverse Contributions

"How many scientists does it take to screw in a lightbulb? Ten—one to screw it in, one to describe it in the paper, and eight to add their names to the author list." It is a well-known joke in the scientific community, and that's for good reason. The process of adding authors can be quite easy, and no one has yet come up with an effective way to prevent the excessive inclusion of names. Therefore, it's often challenging to determine who should be listed as an author and who should not. For example, should the person who conducted all the measurements be included as an author if they didn't work out the results? Or should someone who reviewed the paper and suggested amendments be added to the author list? These are examples of common dilemmas when determining authorship.

Of course, there's no one-size-fits-all answer to these questions. However, in my opinion, there is a rule that strikes the right balance. It comes from COPE (Committee on Publication Ethics) report on scientific authorship, which is based on principles from the International Committee of Medical Journal Editors.

Authorship credit should be based solely on the following criteria:

- Substantial contributions to the conception and design, acquisition of data, or analysis and interpretation of data.
- Drafting the paper or revising it critically for important intellectual content.
- Final approval of the version to be published.

The first two points can be simplified to the question: "Has the person had a significant influence on the research process and the final publication?" This influence could include anything that shapes the outcome of the work. In this context, "significant" can be interpreted as such influencing the final outcome. The third point

4.3 Who Are Your Co-authors?

ensures that all co-authors agree to the publication before submitting the manuscript to the editorial office.

You can also refer to CRediT (Contributor Roles Taxonomy), which was introduced to recognise individual author contributions, reduce authorship disputes, and facilitate collaboration.

What Can Go Wrong?

The most common bad practices when creating a list of authors include:

- Co-authors disagreeing on the readiness to publish.
- Using another person's results without proper acknowledgment.
- Excluding a rightful author.
- Gift authorship (e.g., adding someone simply as a courtesy or "thank you").
- Pressured authorship (e.g., adding someone due to external pressure).
- Ghost authorship (attributing work to a non-existent person).

Any of these practices can lead to the paper being withdrawn and require careful consideration. In the first three cases, a dissatisfied author (or someone who has been unjustly excluded from co-authorship) may contact the editor of the journal. If they can provide evidence of the issue, the paper can be withdrawn.

The next two cases—gift authorship and pressured authorship—occur when someone is added as a co-author despite their minimal contribution. In gift authorship, one co-author may add a person as a courtesy. In pressured authorship, a co-author may be coerced into including someone, perhaps by a superior. If someone contributed in a minor way but didn't meet the criteria for co-authorship, their involvement can be acknowledged in the "acknowledgments" section instead.

Finally, the strangest case: ghost authorship. It's hard to believe, but there have been instances where publishers have discovered that the listed author of a paper doesn't even exist! Some have gone as far as creating fake personas to support and cite their own work. A famous example of a similar flavor is the author F.D.C. Willard. In 1975, Jack Hetherington published a paper in Physical Review Letters. This journal allows first person plural only when a paper has two or more authors. The scientist found a way to avoid revising the work, which was a lenghty process back then. He realized that the cat was hanging around the typewriter and should indeed be treated as the co-author. As a result F.D.C. Willard (Felis Domesticus Chester Willard) is an official author and has his own Wikipedia page.

After highlighting so many bad practices, it's important to focus on good ones. Here's one key piece of advice: the list of authors should be established as early as possible. Ideally, this should be done before writing the publication, and even during the research process. COPE suggests that co-author agreements be signed before starting the writing phase. While this idea sounds good in theory, it's often impractical in real-world situations. It's difficult to imagine a scenario where potential co-authors sit down to draft and sign a contract before even beginning the work on the publication. But who knows? Maybe it could work in some cases.

Sticking to more practical solutions, I recommend that the leading author (usually the corresponding author) sends an email to potential co-authors outlining the planned thematic scope of the paper, the division of responsibilities, and the proposed list of authors, including the planned order. It's important to note that if no one raises objections at this stage, any subsequent changes will require the consent of all co-authors. This approach helps avoid potential conflicts and ensures effective organisation of both the research and writing process.

Takeaway points:

- **The quality of your co-authors is more important than quantity.**
- **Use COPE's recommendations to decide who your co-authors should be, CRediT can be used to recognise individual contributions.**
- **Try to establish all co-authors and their contribution to the project before the paper is written.**

4.4 High-Quality Diagrams and Figures

Using Visuals

Visuals are a powerful way to convey ideas and information effectively. High-quality diagrams and figures are crucial for getting published and publishers pay close attention to their clarity and impact.

Science highlights the power of visual communication. People tend to remember and learn from images more effectively than from text. This form of communication has deep roots, with early civilisations using cave walls as canvases to share knowledge, beliefs, and stories over 40,000 years ago. While text requires processing each letter and word to derive meaning, images are instinctively understood, allowing for quicker and more intuitive comprehension.

Visual storytelling has remained a powerful tool throughout history, and its influence continues to grow in the modern world. The impact of images is undeniable—on social media, visual content is 40 times more engaging, and articles with a picture every 75–100 words receive twice as much attention as those without.

The ease with which we remember visual content over text can be explained by the "picture superiority effect." This phenomenon occurs because images are "double coded" in our minds, triggering both visual and verbal associations, whereas words only invoke one form of meaning. This dual coding makes images more memorable and easier for our brains to process and recall, which is why visuals often leave a stronger impression than written content.

Don't Be an Excel Victim

Visuals are crucial for attracting readers to your paper. Well-designed figures and tables spark interest, encourage readers to engage with the content, and can motivate them to read the entire publication. However, it's not enough to include visuals—only high-quality visuals will elevate a paper from average to exceptional.

High-quality figures lend a professional appearance to your work. Clear, well-structured diagrams make it easier to understand and interpret results. Readers tend

4.4 High-Quality Diagrams and Figures

to associate a polished manuscript with reliable content, increasing their trust in both your results and interpretations. This also applies to editors and reviewers, who often form an initial judgment about the quality of the paper based on a quick evaluation of how the results are presented.

Many authors fall victim to data-processing software like Excel. While these programs are useful for analysing results and creating diagrams, relying on standard templates limits your ability to creatively design how data is presented. As a result, your work may end up looking dull and formulaic—almost like a high school report. While the data is present, it may not be showcased in an engaging way. So, don't be a victim of Excel; take the time to enhance the quality of your visual presentations in your Ph.D. thesis and research papers. As you'll see shortly, it doesn't have to be difficult.

Add Value

When analysing your results, first decide which data should be presented graphically, which should be in tables, and which should be described in the text. If you have only two or three measurement values, it's better not to use a figure—it won't add value or visual appeal. At this stage, focus on the story you want to tell in the paper and plan the sequence of figures and tables accordingly. You can start from an empty file, add the main headings (preferably IMRaD), and then put in the figures to see if they build the story. This will help structure your content more effectively.

The next question is: which graphic form best suits your data? You have several options:

- **Photographs**—These are ideal for helping the reader visualise information, such as a microscopic image.
- **Figures**—These are perfect for displaying large amounts of data clearly.
- **Diagrams**—Use these to highlight key parts of a system or process.
- **Maps**—These are effective for showing spatial relationships between data, either in 2D or 3D form.

You can also draw inspiration from how data is presented in leading journals in your field. Look at how other authors structure their data and create visuals. Journals like Nature and Science offer great examples of top-tier scientific visual communication. Keep in mind, though, that authors in these journals often work with professional graphic designers. Here are some basic points to consider:

- **Use a scale bar**—This helps provide context for measurements or sizes.
- **Explain the meaning of colours and symbols**—Ensure that readers can easily interpret the visuals.
- **Describe the main element in the diagrams**—Clarify what the most important part of the image is.
- **Provide additional explanations in the figure legend**—Offer further details and context for better understanding.

When it comes to creating and editing your figures, there are many software options to choose from: Origin, GIMP, Matlab, DataGraph, Adobe Illustrator, Mathematica, and more. Ultimately, the specific software isn't as important as the approach you take. The key is to create your visuals thoughtfully, ensuring they add value to your overall message. Don't settle for basic charts from Excel—make them your own to enhance the communication of your results. Even a simple process, such as copying a chart from Excel into PowerPoint and adding extra descriptions and elements, can be effective. However, always remember not to manipulate the data to alter or improve your results. It's best to keep copies of the original graphs, files, and metadata, as they may be required by the journal during the peer-review process.

What matters most is that your graphics are visually appealing and enhance the overall quality of the paper. They should be easy to understand at a glance, allowing the reader to quickly grasp the information. The ultimate aim of your visuals is to communicate your message effectively.

Take away points:

- **Visuals are often the quickest way to communicate large amounts of complex information.**
- **Decide whether to use a table, figure, or text depending on the type of data you need to present.**
- **With visuals, you can increase the reader's ability to process information and increase their desire to read on.**

4.5 Cover Letter

Pitch Your Results

With weeks of hard work, you've now written your paper and selected the perfect journal. Sorry to say, it isn't over just yet! Before you send the manuscript, there are a few more steps. The first one: writing a cover letter to the editorial office.

The cover letter is an important document. It's not just about explaining why your paper is worth consideration. It's your opportunity to highlight the new knowledge your research brings to the field. This letter will help the editor make one of two decisions: either send the paper for peer-review or reject it outright through desk rejection.

Don't Risk Desk Rejection

The role of editor is to maintain the quality of the journal by publishing the best possible papers and ensuring timely issue releases. Desk rejection is the first stage of quality screening for submitted papers. The editor, who holds the "desk," must make a quick decision based on an initial assessment of the paper. At this point, they must decide whether to send the paper to reviewers to begin the review process, or to reject it outright without further review.

4.5 Cover Letter

A significant portion of rejections in scientific journals occurs through desk rejection. You might wonder, why is the editor so harsh by not sending the paper for review? To understand this, let's consider the editor's perspective. They are responsible for maintaining the journal's quality, and one of their primary concerns is attracting and managing reviewers. Remember, each paper typically requires 2–3 reviewers, and finding suitable researcher can be very challenging. Sending a mediocre paper to a reviewer not only wastes their time and energy, but it also discourages the reviewer from accepting future assignments. Efficient peer-review process involves carefully selecting reviewers who are well-suited to the paper's subject matter and ensuring they are assigned papers with real potential for publication. Desk rejection helps save reviewers' time and energy, allowing them to focus on papers that are more likely to be published.

Let's put this into perspective with numbers. In prestigious journals, the acceptance rate rarely exceeds 20%. This means the editor can accept only one in every five submissions. Consequently, they can desk-reject about three out of five papers. This strategy helps ensure that the reviewers can focus on the remaining submissions. However, editors are also mindful of journal metrics, such as response times. Accepting papers that show promise but require significant revisions—whether in text or figures—will inevitably delay the process. As a result, editors tend to prioritise papers that demonstrate strong scientific rigour and technical quality.

When deciding the fate of your paper, editors are considering several key qualities:

- **Awareness of the work's value**—Has the author clearly explained what new knowledge the paper brings to the field in the cover letter?
- **Reviewers in mind**—Has the author suggested appropriate reviewers, ideally from outside their institution and from international research centres? More on that in the next section.
- **Bibliography quality**—The bibliography can reveal several key issues. Does the paper include the latest relevant literature on the topic? Are there an excessive number of self-citations (typically, it's recommended not to exceed 10%)? Does the paper incorporate research from a variety of global sources, demonstrating the international significance of the study?
- **Meaningful discussion**—Does the paper go beyond just presenting results to contribute to the development of the field? Has the author provided an in-depth analysis of the results in the discussion section?
- **Clarity and transparency of results**—Are the results presented clearly and transparently, with proper figures and tables?
- **Reproducibility and robustness**—Are the results reproducible, and are trends based on the average of multiple tests? Are error bars provided in figures?
- **Neatness and structure**—Is the paper well-organised, with a clear division into sections and subsections, and formatted according to the journal's guidelines?
- **The core message**—Is the research aim clearly defined and well-justified? Is the main message of the paper explicitly stated? Ultimately, does the paper tell a cohesive story that leads to the core message, or does it simply present a series of results without a clear narrative?

Not every paper must strictly adhere to all the above points, the peer-review process helps to catch and correct these shortcomings. However, if there are too many issues, the safest course of action for the editor might be to reject the paper promptly. This approach helps conserve the time and energy of both the editor and the reviewers, ensuring that their efforts are directed toward more valuable submissions.

The Key Elements of Cover Letter

In addition to the key information required by the journal listed on their website, your cover letter should include the following key elements:

- **Introduction**—Clearly state the title of your paper and the target journal.
- **Significance**—Explain why your content is important for the journal's readers and the broader research field.
- **Research Problem**—Briefly describe the research problem you are addressing.
- **Results and Conclusions**—Provide a concise summary of the main results and conclusions of your work.
- **Publication Status**—Clearly state that the work has not been published elsewhere and is not under consideration by another journal.
- **Additional Details**—Include any other relevant information that will encourage the editor to forward your paper for peer-review.

The cover letter should be concise, ideally fitting onto one A4 page after formatting. Focus on providing the most important information, writing one or two sentences for each key point above. This approach ensures clarity and makes it easier for the editor to assess your submission quickly. If you want to stand out, consider submitting your cover letter on institutional letterhead and include your handwritten signature.

Once your cover letter is ready, all that's left is to send it along with your manuscript and any supplementary files (such as figures, data or code) through the journal's electronic submission system, like Editorial Manager. Phew, another step completed, and you're almost there!

Takeaway points:

- **Your cover letter gives you a chance to speak directly to the editor—make the most of this opportunity.**
- **A good cover letter might make the difference between your paper being desk rejected or sent out to reviewers.**
- **Keep it professional, short and meaningful: one A4 page max.**

4.6 Suggested Reviewers

Step into the Editor's Shoes

When submitting your manuscript, you may be asked to suggest reviewers. Using this opportunity can be a game-changer. When done well, it showcases your professionalism; however, the opposite is also true. Misusing this opportunity can cast doubt on your credibility. Whether or not to suggest reviewers is entirely up to you, but editors place great importance on this choice. It reflects the authors' experience and understanding of their field. So why should you choose to use suggested reviewers? And who would make the ideal choice?

In previous section I explained the issues that editors have with the huge number of reviewers needed to run the journal. They are also very busy. They often juggle editorial responsibilities alongside their scientific research and teaching duties. They regularly monitor the status of papers and must address delays caused by busy reviewers. In some cases, editors may need to contact over a dozen potential reviewers before securing the necessary number of appropriate reviews for a single paper.

In online journal systems like Editorial Manager or Open Journal System, editors maintain a database of reviewers. The quality and diversity of this database are crucial for an efficient peer-review process. However, due to the demands of managing ongoing submissions and editorial tasks, editors often lack the time to actively expand this database.

This is why many editors appreciate suggested reviewers—it spares them considerable effort. Authors, deeply immersed in their research, are often aware of specialists the editor might not have considered. By proposing reviewers, authors not only aid the editor in streamlining the peer-review process but also contribute to expanding the reviewer database. It's an excellent way to demonstrate thoughtfulness and respect for the editor's time.

Help Yourself

However, the editor is not obligated to use the suggested reviewers for your paper. In practice, it's common for one of the suggested reviewers to be invited, while the others are selected from the editor's own database. Even if none of the suggested reviewers are chosen, your contribution still enhances the reviewer database by adding valuable names.

It's always worth suggesting potential reviewers, even if it's not required. Doing so helps the editor and demonstrates your professionalism, showing that you're well-acquainted with your field and confident in engaging with other specialists. So, who should you suggest?

An editor needs reliable reviewers—those who are recognised in their field and can provide insightful feedback. However, this doesn't mean suggesting Nobel Prize winners or other extremely high-profile individuals who are unlikely to have the time to conduct a review. The key is finding a balance: selecting experts who possess substantial knowledge while also being approachable and willing to invest their time and expertise.

Each suggested reviewer should be accompanied by a brief justification. For example, mention that the individual has published an influential paper on the topic or delivered an insightful presentation at a conference. This demonstrates your familiarity with the field and your thoughtful approach to recommending suitable experts.

What Not to Do

A common mistake is suggesting someone with a conflict of interest. While editors typically conduct due diligence, their workload often prevents exhaustive checks. This means they might not always identify potential conflicts, such as the reviewer being from the same institution as the author. Even reviewers from the same country can present a risk of bias, so it's crucial to avoid such situations whenever possible.

Suggested reviewers should not be affiliated with your institution or have co-authored with you in the last five years. Avoid individuals with personal or close professional relationships, such as former students, employees, or supervisors. Ideally, select reviewers from diverse countries, which helps maintain impartiality and credibility in the review process.

If you've co-authored with a suggested reviewer in the past, an editor may view it negatively. It could imply that you believe your work is too weak to be fairly assessed by an independent expert, or that you lack connections to other researchers outside your immediate circle. This might raise concerns about the objectivity and breadth of your network. It's better to suggest reviewers who are outside your close professional ties to demonstrate a broader and more independent perspective.

As with many other aspects of life, first impressions are crucial. For an editor, it's not limited to the paper itself. You have additional chances to stand out. This includes the overall appearance of the paper, the cover letter, and the list of suggested reviewers. It's an opportunity to demonstrate your professionalism, showing that you not only understand your work but also know who will find it relevant. A well-crafted submission creates a strong, professional impression and increases the likelihood that the editor will forward it for review.

Takeaway points:

- **Help yourself by helping the editor and suggest people who can review your paper.**
- **A good reviewer is someone who is familiar with your research area and is not from your close circle of collaborators.**
- **By suggesting reviewers, you show confidence in your paper.**

4.7 Submitting Your Paper

The Job isn't Over Yet

You might think the work on your paper is finished once the final draft is ready, but it's not that simple. The work isn't truly complete until your paper is approved and published in a journal. Even if you have the cover letter and the list of suggested reviewers you are still not there. There's one last crucial step before submission: packaging your manuscript and presenting it in an attractive, professional manner to both editors and reviewers.

A significant number of papers are rejected early in the submission process for failing to meet basic technical requirements, as we discussed in Sect. 4.1. Many of these papers are desk-rejected before they even reach the reviewers. Naturally, you want to avoid this, so let's review the key points to check before submitting your manuscript to a journal.

Your Checklist

The following list is a great way to check the overall quality of your paper before submitting it. I'll break it down piece by piece, starting with the title:

- Is the title concise and engaging? Does it grab the reader's attention and stand out among other papers in the journal? A strong title is key to capturing the reader's interest, as explained in Sect. 3.3.

Next, the abstract. A well-crafted abstract should pull the reader in, summarising the story of your paper. If you need guidance on writing a great abstract, refer back to Sect. 3.4. In short, here's what to focus on:

- Does it clearly explain why the research was conducted, why the results matter, and what the main conclusions of the work are?

Moving on to the Introduction (covered in more detail in Sect. 3.5). It should describe the problem, why it's important, what others have contributed to the current state of knowledge, what they've shown, what your work adds to this, and your hypothesis. Here's a quick checklist for the Introduction:

- Does it explain the broader context of the work and justify the need for research?
- Have you cited the most important works from other authors in the field?

Evaluate the entire paper. It should be presented logically, with each section forming a cohesive whole. Each paragraph should focus on a single thought or conclusion, without mixing multiple ideas. Other considerations include:

- Do you follow the IMRaD structure (Introduction, Methods, Results, and Discussion)?
- Does the title and order of sections and subsections maintain clarity and flow?
- Is each paragraph long enough to be substantive but not too lengthy, ideally no longer than a typed page?

The next few points in the checklist focus on data presentation and the quality of figures and tables:

- Do all tables and figures have captions?
- Are all tables and figures mentioned in the text?

- Are the figure elements and fonts large enough to remain legible when reduced to column width in the paper?
- Are all figures and tables numbered in the order they appear in the text?
- If previously published figures are used, do you have permission from the publisher to reproduce them?

Based on the figures, the editor and reviewers can quickly form an opinion about the quality of the entire paper. It's worth putting effort into effectively presenting the data in a way that is easy for the reader to understand. Most editors also look at the Discussion section before sending the text for review:

- In the Discussion section, do you just present the results, or are you taking the time to interpret them in the context of the current state of knowledge?

In a good Discussion, the authors go beyond merely presenting the results; they explore what those results could mean within the broader context of the field. It's important not just to propose a theory, but to explain why that theory should be applied. Don't just provide information—draw meaningful conclusions.

It's also important to review the bibliography:

- Is the number and quality of references adequate?
- Are there too many self-citations in the references? Typical limit is 10%.
- Does the list include full bibliographic details with complete titles?

A well-curated bibliography confirms the author's competence and signals the value of the paper.

As discussed earlier, the cover letter provides an opportunity to directly address the editor and explain why the paper should be published in their journal. Key considerations include:

- Is the cover letter tailored to the editor of the journal?
- Are the editor's details in the cover letter up to date?

Finally, consider these formatting issues:

- Have you removed unnecessary abbreviations and clarified ambiguous terms?
- Except for common mathematical symbols, have you explained all the symbols used in your work?

This is a relatively lengthy list, but it's worth going through when submitting your paper. While these points generally apply to all journals, be sure to review each journal's specific requirements. Before submitting, check their "guide for authors" page to ensure you meet any additional formatting or submission guidelines they may have.

Submission Process

The submission process itself typically takes around 1–3 hours to complete. It will involve a multi-stage questionnaire regarding your submission. Don't be surprised if you are asked to fill in fields for the abstract, title, and keywords, even though they are already included in your file. The system requires them separately for the purpose of inviting appropriate reviewers. Keywords help identify the right candidates, while the title and abstract are presented before a potential reviewer agrees to review your work. Stay calm and continue filling out the submission form—while it may feel a bit over-engineered, trust that everything is there for a reason.

Finally, you will be asked to attach the PDF of your paper. The process may vary depending on the submission system used by the journal, but in most cases, there is one final step that can be a bit confusing. After submitting the file, it may appear as though the submission is complete, but it's not just yet, so be patient. The system will generate a final PDF that includes all the information you provided, along with attachments such as the cover letter and the paper itself. You will need to download this file and carefully check that it accurately represents your submission before finalising the process.

And that's when the magic "submit" button finally appears. This is the moment when your paper is officially considered submitted. You can confirm this by checking the paper's status in the system. It's worth visiting that page regularly to monitor the progress of your submission. Most systems will indicate when the paper has been sent to reviewers, which is a crucial sign that your paper has not been desk rejected.

Takeaway points:

- **Your paper draft isn't finished until it's finally submitted to a journal.**
- **This stage requires your full attention; don't risk desk rejection by making simple formatting mistakes.**
- **The "guide for authors" section on the journal website is your friend.**

4.8 Reply to Reviewers Only Once

What's at Stake?

Revising your paper is inevitable, and it's likely you'll be asked for either a minor or major revision. Don't worry—if you are asked for the review, the main message is that you are still in the game! The key is ensuring that your first revision is also the last. Multiple rounds of revisions increase the risk of an editor becoming frustrated with the lengthy review process and, ultimately, rejecting your paper. Aim to address all feedback thoroughly in the first round to avoid prolonging the process and getting the negative outcome.

A paper review often seems more daunting upon first reading. My first piece of advice is to give it some time. Especially when asked to make significant revisions, it's helpful to step away for a day or two before revisiting it with a clearer perspective. And going to a park or a hike is the best way to make it look better. Remember, if your paper hasn't been rejected, you're already in a strong position! It has been pre-selected as suitable for the journal, and now you simply need to address the reviewers' suggestions. Instead of seeing revisions as a setback, view them as an opportunity to enhance your paper's quality. After all, the reviewer is an expert who has invested time in your work and offered constructive feedback to help you bring it to the journal quality.

Don't Leave Anything On the Table

Your response should be crafted to make it as easy as possible for the reviewers and editor to accept the revised paper. The most important point: respond to all the reviewers' points. It may be tempting to skip inconvenient comments. It doesn't matter what's the reason: you don't want to do additional research, you disagree with a suggestion, you think the reviewer asks stupid question... Just please never leave any points unaddressed. If you do - editors and reviewers will notice. Even if you disagree with a comment, it's a much better strategy to explain why you haven't made the suggested change.

Leaving comments unaddressed is one of the worst mistakes you can make. Imagine the editor in a situation where only 8 out of 10 questions are responded to. What should they do with that? It's clear that the paper cannot be accepted in this form. Sending it to reviewers with only partial responses will only frustrate them. Reviewers are likely to feel that their valuable feedback has been ignored. In the best-case scenario, the editor will return the paper to you to address the remaining comments. But they may not be that patient... With limited options, rejection becomes more likely. The editor probably has other papers in the pipeline, and the one with authors who are most cooperative is more likely to be accepted.

Three Steps

There are three key steps that will significantly increase your chances of an editor accepting your revised paper:

- Present the amendments in a structured and transparent way.
- Clearly distinguish the revised text in the paper.
- Include a letter to the editor.

The first point concerns how you present your amendments. The most effective approach is to create a separate "Response to Reviewers" document with a table that includes the following columns:

- The number of the reviewer's comment.
- The original comment.
- The page number the comment refers to.
- Your response.

By clearly distinguishing, i.e., highlighting, the revised text in the paper, you'll make the editor's and reviewer's tasks much easier. You also direct their attention solely to the added or revised content. This subtly shifts focus onto the amendments and reduces the likelihood of new comments that weren't part of the original review.

The third point is to attach a letter to the editor. In this letter, you should thank the editor for the reviews and explain how you have addressed the suggested revisions in your response. It could be written like this:

4.8 Reply to Reviewers Only Once

"Dear Editor,

Thank you for your email dated [insert date], enclosing the reviewers' comments. We are grateful to the reviewers for their time and constructive feedback. We have carefully considered their comments and revised the manuscript accordingly. Our responses to each point are provided in the attached "Response to Reviewers" document. Changes to the text are highlighted in red in the revised manuscript.

We hope the revised version meets the journal's requirements and is now suitable for publication. We look forward to your response."

If you follow the three points above and respond to all the comments, your revised paper should be well received. While this process requires additional effort, remember that it also presents an opportunity to enhance the quality of the paper.

Takeaway points:

- **Your priority is to make sure you don't need to revise the paper more than once**.
- **Don't annoy the editor; reply to ALL reviewers' comments.**
- **Make sure you attach both a "Response to Reviewers" and a letter to the editor with your revised manuscript.**

Chapter 5
Master Conferences and Presentations

> Conference is a chance to meet the top minds, and turn your research from a whisper into a roar.

5.1 Preparing for the Conference

Don't Miss Your Chance

Attending conferences can give boost to your career and stimulate your personal growth. You can showcase your experience and learn from others. It's a place where you can make contacts that last through your professional career and build your position in the research community. All of this in just a few days—who wouldn't want that wonderful chance?

I didn't.

In the early stages of my research career, I struggled to grasp the true purpose of conferences. Here's my "wall of shame," a list of mistakes I made:

- Focused solely on my presentation.
- Overloaded my slides with so many charts and details that no one on Earth could possibly follow them.
- Didn't really ask for feedback.
- Didn't work on my presentation skills, reasoning that I was a scientist, not a performer.
- Paid attention only to presentations closely related to my field.
- Spent coffee breaks glued to my phone or hiding behind my laptop.

Preparing for a conference involves much more than just creating a presentation. While the possibility to present your research is an essential component, it's only one of many opportunities a conference can offer. Attending solely to deliver your presentation is akin to visiting Disneyland for just one attraction—it hardly makes the most of the experience. When visiting Disneyland, you'd typically buy a ticket, book a flight, arrange accommodation, and explore as many attractions as time allows.

Adopting a similar mindset for a conference ensures you can fully capitalise on the opportunities available.

The Conference Starts Long Before You Step Onto the Plane

You want to use the conference time in the most efficient way. You're spending days packing and flying to the other side of the world—make sure it's worth it. So how to make your next conference really meaningful for your development? It will require a little preparation upfront.

You should browse the conference programme and get acquainted with everything that's going on. This will help you see if you can participate in all sessions or if you need to prioritise the most important ones. In big conferences you will have many parallel sessions and many difficult choices to make. This is a good time to decide which presentations are most important from your perspective and make a note of them. At some conferences, there's more to experience than just lectures. Workshops and tutorials may also be available, offering excellent opportunities to gain practical knowledge. Workshops specifically designed for Ph.D. students can be particularly valuable. Some conference events may require prior registration, which gives additional argument to do it in advance.

After completing that part of the preparation, you should have your own personalised conference agenda. With it in hand, you'll be able to navigate the event smoothly without having to shuffle through the official timetables or search for the right room for the next session. You can also add additional activities to your agenda. Let it be dinner with someone or visit to local laboratory. You can do it on paper or use your regular Outlook/Google calendar just as it is normal day of work. I find this approach especially useful, as those calendars give automatic reminders before each session starts.

If you plan to visit a laboratory or research group at a nearby university, ensure that the appointment doesn't conflict with the conference schedule. It's much more courteous to arrange a meeting in advance. In some cases, you may discover that meeting during the conference isn't feasible, and it may be worth extending your stay by a day or two to accommodate the visit. If you want to arrange the visit contacting the head of the laboratory might not be the best option. It's usually more effective to reach out to someone in a position like yours. They often have more time and, like you, are looking to expand their network. It's also worth checking if you have any mutual connections—LinkedIn can do that for you. If so, you can request a quick introduction from the person who knows both of you. If you can facilitate this, an in-person visit is almost guaranteed.

Online Conferences

Virtual events present an interesting alternative to traditional conferences or add a valuable complement to them. They offer several advantages: they are more affordable, eliminate travel time, and provide opportunities for those who might otherwise be unable to attend, such as young parents. But as the name suggests, virtual conferences only offer a fraction of the full in-person experience.

If circumstances allow, I recommend attending physical conferences regularly—ideally once a year, or more frequently if possible. Virtual conferences can serve as a worthwhile addition to these in-person events.

You can take proactive steps to maximise the value of your virtual conference attendance:

- **Block time slots**—It can be tempting to juggle virtual conference participation with other responsibilities. From my experience I can tell, that whenever I was trying to do two things at the same time, I was usually not doing any of them properly. My advice would be then to ensure you've allocated dedicated time to fully engage in the online event without too many interruptions.
- **Prepare your environment**—Time isn't the only resource to secure—you'll need a suitable, distraction-free space. If you work in an open-plan office, attending the conference from home might be a better option. Also, ensure you have snacks and beverages on hand, especially for longer sessions.
- **Familiarise yourself with the platform**—The rise of virtual events during the Covid-19 pandemic has introduced a plethora of tools and platforms. Make sure you're comfortable with the one being used for the event. Some platforms may require you to set up a profile, including a bio, photo, and links. Preparing this in advance allows you to focus on the sessions during the event.
- **Follow up promptly**—If you plan to connect with speakers or participants, do so immediately after a session. Mention that you attended their session and would like to connect. We're more likely to forget people met during virtual events, so timeliness is key.

Takeaway points:

- **Conferences offer incredible opportunities to advance your career.**
- **Alongside official conference agenda you should prepare your own schedule.**
- **For virtual conferences, ensure you've secured both time and a distraction-free space to fully engage.**

5.2 Making Meaningful Connections at the Conference

OMG, I Will Meet Prof. Smith!

Here's a scenario: you're preparing for a conference (as discussed in Sect. 5.1) when you realise that Prof. Smith is also presenting at the same event. Can you imagine? Naturally, you'll want to connect with her. She's the star! Having that contact could prove invaluable in the future. But how do you go about preparing to interact with such a person? My advice is again, to be prepared.

Do you know her research well? Take the time to review her latest papers and any recent work she published. Familiarise yourself with the key ideas, contributions, and any current projects she's involved in. This will not only allow you to engage in meaningful conversation but also help you ask insightful questions that demonstrate your genuine interest in her work.

Take the opportunity to make a strong first impression. Her presentation is likely to leave you with some questions—so ask them in the Q&A session. Asking thoughtful questions can serve as a great starting point for further conversation. If you'd prefer a more personal exchange, consider bringing your questions to a coffee or lunch break.

If you hadn't had a chance to attend her talk, you will need to find the time and space during the event to approach her. Ideally, you want to engage when she is not distracted by other people. If she is rushing between the session, she might not have the time for a meaningful conversation. Look for quieter moments where you can have a more in-depth discussion. If the opportunity arises, don't hesitate—especially with busy individuals like Prof. Smith who is often approached by many. That might be the only chance you will get.

During the conversation, ask the questions you've prepared and then let the conversation unfold naturally. You can also briefly mention what you do. At the end of the conversation, thank her for her time. It's a good sign if you exchange business cards, particularly if she offers it first—this indicates he is interested in staying in touch. If you want to increase your chances of further contact, you can offer to send valuable resources after the meeting. If nothing specific comes to your mind, you can always ask if it would be alright to get in touch later with any questions related to your research.

Active Listening

Have you heard the term "active listening"? It refers to fully engaging in the process of listening. While it's easy to be an "active speaker"—the world is full of them—achieving the same level of engagement as a listener is much more challenging.

As an active listener, you use all your attention to ensure you fully understand your conversation partner. It's a skill that can be developed. Here are some tips to help you make the most of important conversations, like the one with Prof. Smith:

- **Maintain eye contact with the speaker**—Looking around can be interpreted as a lack of interest in the conversation.
- **Face the speaker**—This isn't just about your eyes but your whole body language. If you're speaking with someone, make sure your chest is directed towards that person. If only your head is turned towards the speaker, you may appear as if you're just passing by and briefly joining the conversation.
- **Rephrase**—You can summarise what you understood by saying "So what you mean is...". That's a great sign of your interest and a good way to make sure that what you understood is what the speaker intended to say.
- **Phone destroys conversation**—Swiping your phone when someone speaks is an extreme situation—it looks like you are not interested at all. But even slight presence of a phone is harmful for the conversation. Imagine someone keeps the phone off in the hand during the conversation. It can be still viewed as a distraction. So, the rule is simple—if you want to have a good conversation—hide your phone.

- **Ask open ended questions**—Open ended questions are great way to lift the conversation to higher level. You let the speaker respond in the natural way, so the conversation feels comfortable to both sides.
- **Turn on your feelings**—Try to feel the story with your own feelings. If the speaker is telling about some exciting experiment, you can imagine being part of it. You can share what you feel by saying "that must have been awesome, I wish I participated in an experiment like that".

These are just some of the possibilities of how to elevate your active listening skills. The best part is that each of them can be tried during your next conversation, and you will immediately see the result. You can also ask friend to try different variants of the conversation. For example—you can try listening to your friend for three minutes with your phone in hand and then for another three minutes without the phone. You can then reflect how it felt for both sides of the conversation.

Active listening is an excellent way to make the speaker feel more comfortable. It also enhances the quality of the conversation, turning it into a more engaging and enjoyable experience.

Effective Follow-Up

If you'd like to keep in touch, the easiest way to do so is by connecting on social media. In a professional setting, LinkedIn is your best bet. Be sure to send the connection request within 48 h of the meeting, while the interaction is still fresh in Prof. Smith's mind. This is crucial, as it will help her immediately recall who you are.

If the meeting was particularly valuable to you, consider sending a follow-up email. This should also be done within 48 h of the interaction. In the email, express gratitude for Prof. Smith's time, and if appropriate, include some relevant materials that could be of interest to her. This might include a link to your own work, research, or any valuable resources you believe she would appreciate.

Once you've established this connection, you can gradually nurture the relationship over time. On LinkedIn, for instance, you can engage with Prof. Smith's posts, leave thoughtful comments, and ask questions. This helps maintain a interaction without feeling intrusive.

Now that the ice is broken, future interactions at conferences and other events will feel much more comfortable. These relationships could open doors to new opportunities, collaborations, or even mentorship. But always remember: it all begins with that initial conversation. Make sure it happens.

And as a final note, everything I've mentioned in this section applies to professors of all names, not just "Prof. Smith"!

Takeaway points:

- **Check before the conference who is attending and find people you would like to interact with.**
- **Have some questions prepared and then let the conversation take its course.**
- **You should follow up within 48 h after the initial meeting.**

5.3 Five Steps to an Impactful Presentation

Steps 1 and 2: From Concept to Rough Design

Here are the five steps I typically follow to ensure my presentation is as polished as possible. I'll cover how to plan it, refine the flow, and prepare yourself for the big day. So, let's dive in.

Conceptualisation—The first step is to plan your presentation content in a way that feels comfortable to you and ensures the information is engaging for your audience. Start by brainstorming with a blank sheet of paper, focusing on crafting a catchy title and a concise abstract that effectively reflects the essence of your presentation.

At this stage, your goal is simply to define the story you want to tell. It's best to begin this process as early as possible. Many conference organisers will require the title and abstract several months in advance as part of their registration process.

Presentation outline—About a month before your presentation, it's time to start working on the proper outline. Before creating the slide deck, I recommend returning to a blank sheet of paper. Based on your abstract, you can define the main sections of your presentation, such as: introduction, historical context, research method, key conclusions with examples, future outlook, and a summary. For each section, jot down the key points you want to convey. Once you've completed this, review the outline to ensure your story is coherent and engaging.

Now you're ready to move to PowerPoint (or another software of your choice). Start by creating separate sections in the presentation (a very useful feature in PowerPoint) based on your handwritten plan. Then, dive into each section and loosely design slides with the content you know should be included. Don't worry about the design at this stage—the focus should be on getting the overall view of the presentation. You can add bullet points, key plots, and diagrams. If you don't have the satisfactory version of the content, just put some placeholder. By the end of this step, you'll have rough-looking slides, and the overall presentation will tell a coherent story.

Steps 3 and 4: Polishing Slides and Key Insights

Visuals and flow—At this stage, you can begin working on the visual elements of your presentation. It's helpful to focus on each section individually, breaking the process down into smaller, manageable pieces. Once you're satisfied with the visual aspects, shift your attention to what you want to say for each slide.

Simply talking through your presentation to yourself might not work for everyone. My approach is to print out a three-slide layout with space for notes on the right. As you work on the notes, you'll likely spot areas that need adjustment. This is precisely the purpose of this step—to refine the flow of your narrative and smooth out transitions between sections of the presentation.

Key insights—Ideally, you'll have your slides ready a week before the presentation. Now, it's time to condense your speaker notes. You don't want to be one of those presenters who reads every word from their notes. Instead, aim to rewrite them into a much more compact version.

At this point, you should have identified two to four key insights for each slide. This balance allows you to speak naturally in your own words while ensuring you cover everything necessary for the overall flow of your presentation.

I once had the opportunity to observe a speech delivered by the Deputy Chief of Mission at the UK embassy (the official title for the Vice-Ambassador). His speech was outstanding. Being in close proximity to him, I noticed how professionally he was prepared. What surprised me was that all he had were small cards, each with just two or three key points. He followed this process: he would quickly glance at the card, speak a few sentences, then switch to the next one. With a deck of 10 little cards, he delivered a remarkable speech.

You can adopt a similar approach for research conferences, whether using PowerPoint notes, handwritten notes, or preparing small cards for each slide, to speak like the Vice-Ambassador!

Step 5: Practice, Practice, Practice

Keep doing that—A few days before the presentation, you'll likely start feeling a surge of adrenaline. This is the perfect time to run through the entire slide deck. If you can find time to rehearse multiple times, you'll notice that you start relying less on your notes. You may even reach a point where you can deliver the entire presentation without them. That's when you've internalised the key points and could potentially present without any notes. However, I would still recommend having your cards or notes with you for the final presentation, even if just in your pocket. Sometimes, the mere knowledge that you have a backup plan can help you relax enough that you don't need to use it.

If you're someone who gets extremely stressed before a talk like most of us consider doing a mock presentation in front of a friendly audience. While it may not replicate the exact conference environment, presenting your content to others is always valuable practice.

And there you have it: five simple steps to make your presentation the best it can be. When you break it down this way, the process seems less daunting. You also work in the right order—starting with the big picture, then moving to the details and final preparation. Give yourself the time and space to prepare, so you'll know that when you step out to present, you'll give the best performance possible.

Takeaway points:

- **Before creating the first slide, visualise your presentation on paper.**
- **Keep working on slides until you form a clear story.**
- **Write down 2–4 key insights for each slide.**

5.4 Three Constraints That Shape Your Presentation

Constraint 1: Message

When preparing for your presentation, your mind will likely be buzzing with ideas. And it's no surprise—you finally have the chance to present the results of your hard work. Naturally, you'll want to share all the details. However, presentation time is usually limited, so you'll face a difficult decision: what to highlight, what to leave out, what's truly important, and what isn't. It's incredibly challenging to view your work with objectivity and trim down details without feeling attached to them. After all, each piece of information contributes to the larger picture of your achievements. So, how can you strike the right balance, ensuring that you convey your full story without overwhelming or boring the audience?

Consider what the most important message of your presentation is—what key knowledge do you want to impart to your audience? In Sect. 3.2, I discussed the core message in a research paper, and the same principle applies to presentations. By the time you're preparing your slides, you should have already defined your central theme.

If you're following the steps outlined in the previous section, it might be helpful to define your core message in the conceptualization stage. Write it down and evaluate whether it's concise and engaging. If your message feels convoluted, it could be a sign that you're trying to convey too many ideas in one presentation. The core message will also serve as a compass helping in selection of slides. For each slide you can ask the same question—is it connected with the presentation core message?

Constraint 2: Time

Conference chairs and moderators are typically strict about adhering to time limits. Exceeding your allocated time will almost certain result in being rushed, which can undermine the impression your presentation makes on the audience. Moreover, going over time reduces the opportunity for questions.

Consider this: top scientists in your field have taken time out of their busy schedules to engage with your research. This is a unique opportunity to gain fresh perspectives and valuable insights. By overrunning your presentation, you risk forfeiting this golden chance to connect and learn from their expertise.

How can you keep track of time during your presentation? A test run is essential. Ideally, this should be conducted in front of an audience. Practicing alone won't fully capture the experience of speaking to a room full of people, which is why I strongly recommend organising a mock session. Keep in mind that the dynamics of your presentation may shift when stress is involved. Some people, like me, tend to speak faster under pressure, while others slow down. Knowing your tendencies is key.

I find what I call "checkpoints" especially helpful for tracking time during a presentation. Here's how to use them effectively:

5.4 Three Constraints That Shape Your Presentation

- Identify 2–3 key moments in your presentation to serve as checkpoints. For example, checkpoint 1 could be the end of your introduction, and checkpoint 2 might occur midway through presenting your results. Ideally, these checkpoints should divide your presentation into roughly equal segments. If your talk is 15 min long, good checkpoints would be around minutes 5 and 10.
- During your mock session, measure the time it takes to reach each checkpoint. This will provide an excellent reference for the actual presentation. At each checkpoint, you can assess your pace: are you ahead of schedule with some buffer time, or are you falling behind and need to speed up?

If you find yourself ahead of schedule, you're in an excellent position. You can slow down slightly to give the audience more time to absorb the information. Another effective strategy is to harness the power of the pause—there's no rule that requires you to speak continuously. Pausing at key moments allows your audience to reflect on and process what you've just shared, enhancing their understanding and engagement.

If you find yourself running behind schedule, it can feel stressful, but don't be too hard on yourself—this happens to everyone. The important thing is to handle the situation proactively. In most cases, skipping content is a much better option than rushing through the remainder of your presentation. Skipping a 2–3 slides and maintaining your normal pace means only those slides are omitted. Keep in mind that your audience doesn't know what you originally planned to present. In fact, many won't even notice the skipped slides, allowing you to manage the situation smoothly. On the other hand, rushing through all the remaining slides can overwhelm your audience and significantly reduce their ability to follow and absorb your content. This approach often causes more harm, leaving your audience not able to follow faster pace and, as a result—missing more than 2–3 slides.

Constraint 3: Information

There's another often-overlooked constraint: the amount of information you can effectively include on each slide. We've all seen presentations where a speaker tries to cram multiple charts and blocks of text onto a single slide. This approach often fails. The audience will get lost in the overload of information and likely won't grasp the core message.

Think of it this way: the goal of the presentation isn't to showcase all your results. Instead, it's to spark the audience's interest in your research. If you succeed in doing that, they'll be motivated to explore your publications, where all the detailed results are presented.

There is a physical limit to how much information the audience can absorb in a short time. Once you exceed that limit, you risk losing them. It's worth considering this limit for every slide. If you've followed the structure outlined in the previous section, you can assess this during step 4, when you note down the key insights for each slide. If you find there are more than 3 of them, consider splitting the slide into two.

Takeaway points:

- **Make sure you know your presentation's core message.**
- **Mockup presentations and checkpoints can help you in managing time with confidence.**
- **Each slide should have no more than 2–3 key insights.**

5.5 Getting the Audience Crazy About Your Content

Storytelling

I've spent some time discussing the challenges of presenting—the limitations and roadblocks. Now, let's focus on a more positive side. When it comes to pushing boundaries and overcoming limitations, you have some hidden superpowers. These are simple techniques that can help you communicate more effectively. You're probably already familiar with them, as you'll often hear these methods used in radio, TV, and podcasts. In the context of scientific presentations, they may require a bit more effort, but the results will be well worth it.

You've probably seen some TED Talks, and one of my personal favourites is Hans Rosling's presentation, "The Magic Washing Machine". In this talk, he skilfully used data to show the progress in the fight against poverty in an engaging and accessible way. The title alone grabs attention, and the opening sets the stage perfectly. He shared a personal story from his family home when his mother was using the washing machine for the first time. He used this real-life analogy to connect with the audience. To enhance the story, he even had the washing machine on stage and was reproducing how he remembers this moment. This relatable introduction immediately draws people in, giving Hans the opportunity to guide them through the rest of his presentation with ease and authority.

Hans effectively used the well-known storytelling technique. Our minds process information more easily when it's paired with images, and storytelling taps into this. You're probably familiar with this technique from your own experiences—it's something you've likely encountered many times. Can you think of any examples now? I'm sure you can, because storytelling is used from our very first days. Parents read stories to their kids to share the very first life lessons.

If you don't have a story prepared for your presentation, don't worry—you can always create one. Having a real-life story is a great asset but not having it cannot block you from using this powerful technique. You can always shape your point into the story.

For example, when I applied for a research grant to develop a safety system for compressors, I used storytelling when presenting my case to the panel of experts. I began the meeting with a question: "Imagine you're the managers of a large chemical factory. At some point, the compressor—which is the heart of the entire plant—breaks down, and you must stop the production. Since you're waiting for a custom-made

5.5 Getting the Audience Crazy About Your Content

compressor, you lose income for several weeks while dozens of employees still need their salaries. Wouldn't you want to protect yourself against such a situation?"

Instead of just presenting the facts and pleading my case, I created a vivid image that placed the experts in a real-world scenario. This helped them consider the gravity of the problem and made them more invested in my solution. By engaging them emotionally with a practical situation, I set up the perfect segue into the scientific aspects of my solution, making the discussion more impactful.

How to Present Numbers?

As scientists, we often have to present numbers, but these can be difficult for an audience to fully grasp—especially if they're not familiar with the specific parameters or physical values we're discussing. In such cases, it's important to provide a point of reference that makes the data more relatable.

I once read a book on tribology, the science of friction, and the author used a brilliant example to illustrate its significance. On the first page, he wrote, "Imagine that all the engines in the world had their coefficient of friction lowered by just 0.01." At first, this may sound like a simple and achievable goal, but why is it so important? The author then went on to explain that the energy savings from this small reduction would be enough to power several European countries. By tying the concept to something real and tangible, he made the science not only understandable but also memorable. Even though I can't recall the exact book it came from, that example has stuck with me ever since.

I was once invited to speak at a science festival for high-school students. My presentation was scheduled right before a lecture by Dalibor Carbol, renowned for his Himalayan adventures on paraglider. The challenge I faced was explaining how a paraglider flies, and, even more complex, explaining atmospheric pressure to a group of teenagers.

I started by giving them a number: 101,325 Pascals. But I quickly realised that this number didn't mean anything to them. I could have said 83,924,754 Pascals, and it would have made no difference—without any reference, the number was just abstract. So, I turned to something more relatable: I referred to the force that atmospheric pressure exerts on each of us and presented that it is equivalent to seven punches from Vitali Klitschko or 20 footballs kicked by Cristiano Ronaldo. Immediately, the concept clicked in their minds. They understood the magnitude of the force pressing down on them.

This led to a natural follow-up question: why don't we feel this force? This allowed me to introduce the concept of vacuums, and I demonstrated a well-known experiment with an egg being sucked into a bottle with a candle inside of it. A few simple, relatable examples helped me explain Bernoulli's law and the principles behind how a wing works. The students, far from being disinterested, kept asking insightful questions.

It was a great reminder that when we take the time to make complex ideas relatable, we can engage people of all ages—even those who might seem far removed from the topic at hand.

Opening and Ending

When I help Ph.D. students prepare for presentations, I always emphasise the importance of a strong opening and conclusion. Just as an exciting introduction and ending are vital in fiction book, they are equally crucial for your non-fiction presentation.

Opening—At the start, it's natural to feel stressed. That's why it's worth preparing the first introductory sentences in advance. If you can begin confidently, the rest of the presentation will likely follow more smoothly. Be sure to start by thanking the person who introduced you and offering appreciation to your co-authors. If you have a clear main message, use it as a strong foundation for your opening.

A strong, confident voice in your opening lines is essential to give the audience the impression that you are well-prepared. It may sound odd, but your mind is also listening to you. When you speak with self-assurance, your body will naturally respond, helping you relax and feel more at ease.

Closing—At the end, aim to summarise your key points with three to four clear conclusions, ideally presented as short bullet points. To reinforce your message, consider reusing a graphic from earlier slides in your presentation. It can be scaled down version of a figure, with the goal of helping the audience make the connection and retain the information. Finally, don't forget to thank your listeners and provide your contact details as you wrap up.

Q&A—After delivering your presentation, it's a nice habit to thank each person who asks a question. If you're unsure about what's being asked, it's perfectly acceptable to try to re-phrase the question by saying, "If I understood the question correctly...". Always answer as honestly as possible, and don't hesitate to admit if you don't know something. In such cases, simply acknowledge that you'll need to check the information and offer to follow up with an answer via email.

I hope you found some of the presented techniques useful. I sometimes hear the argument that such "tricks" aren't necessary for scientists—that at a conference, the focus should be on the content, and it's the audience's responsibility to stay engaged. However, there are a few counterarguments to this perspective.

First, a conference is an intense intellectual experience, so it's important to help your audience stay aligned with your key messages. They are dedicating their time to listen to you, so it's important to show that you value their attention and participation.

Second, not everyone in the audience may be from your specific field, so it's essential to make your points accessible to a broader audience.

Finally, I'd like to quote Richard Feynman: *"If you can't explain something in simple terms, you don't understand it"*. This is a reminder that simplifying complex ideas helps both you and your audience gain a deeper understanding of the discussed issue.

Takeaway points:

- **Storytelling is a powerful tool to deliver your message and keep audience engaged.**
- **Avoid presenting abstract numbers—try to relate them to something your audience will understand.**
- **Work on strong opening and ending for any presentation.**

5.6 Why Body Language Really Matters

The Audience Is Watching You—Are You Watching Them?

Not everyone places the same importance on body language, just as with other techniques for engaging the audience. But let me convince you by describing an intriguing experiment. A renowned psychologist once set out to explore how a speaker's appearance affects the audience's reception of their content. To test this, he sent Dr Myron L. Fox to deliver a presentation at a conference. In reality, Dr Fox was Michael J. Fox, the famous American actor known for his role in Back to the Future.

Dr Fox had a great presentation prepared, excellent diction, and, crucially, strong body language skills. However, he was given a particularly challenging task: to present a talk on "Mathematical Game Theory Applied to Physical Education". Of course, this was a completely fictional and intentionally flawed topic. Dr Fox had an hour-long presentation that made absolutely no sense. His audience? Ten industry experts. It was, without a doubt, a risky mission.

Michael Fox was convinced that his audience would tear him apart. Yet, his presentation was surprisingly well received, generating numerous questions that required him to channel his inner politician—offering plenty of words without much substance. Afterward, all ten experts described the speech as inspiring, and nine of them said it was presented in an accessible manner. The experiment was repeated in various contexts, and it is now widely known as the "Dr Fox Effect".

I'm not sharing this story to suggest that you should find an actor to stand in for you, nor am I implying that the content of your presentation doesn't matter. Rather, I hope this example highlights how crucial body language and presence are in ensuring your message reaches your audience effectively.

Few people have the ability to give a presentation without feeling some stress. Even seasoned speakers often experience a rush of nerves just before stepping onto the stage. A mild dose of stress can be motivating and help you perform at your best. However, it's important not to let that stress paralyse you. I have a few tricks that help me stay mindful of my presence on stage, and I am happy to share them with you.

My first piece of advice, which can lead to immediate improvement, is to work on your eye contact. We've all seen a presenter who relentlessly stares at the slides. I can't help but wonder what they're looking at—after all, they know the material well! This often suggests the speaker hasn't considered their body language during the presentation. It may also be a sign that stress is pushing them to avoid eye contact with the audience. In any case, it negatively impacts the audience's perception of the content. Aim to maintain calm, purposeful eye contact with your audience throughout your presentation.

If that sounds hard, I have another tip that might help. Focus on a few selected individuals in the audience. Somewhere out there, there's likely someone who looks friendly, maybe even smiling. These are your allies. Imagine you're speaking only to them. You've probably had many conversations with a friend or colleague about your experiments, so treat this as a similar situation. To avoid making your "ally" uncomfortable, simply shift your focus to another friendly face, or broaden your gaze to the rest of the audience.

What to Do With Your Body?

I'm sure you've experienced that awkward moment during a presentation: you're in the flow, everything is going wonderfully, and then suddenly, you become acutely aware of your body. Are you standing in an odd way? Where do you put your hands? This brief moment of uncertainty can quickly undermine your confidence. Here are some general tips on how to maintain calm presence during presentation.

Your hands, in particular, are a powerful form of communication. They can convey a lot, often without you intending it. Excessive movement can distract from your message, while putting your hands in your pockets might be perceived by some as a sign of disrespect. Crossing your arms can create an impression of being closed off or distant, as though you're shutting the audience out.

So, what should you do with your hands? Everyone has their own natural style, so it's worth taking a moment to observe yours. But it is recommended to limit your hand gestures to the invisible box located in front of your belly and chest. Ask a colleague to record you while you speak. If your hand movements appear unnatural, you can connect them in the form of triangle, sometimes referred to as the "steeple". Alternatively, you can hold an object, such as a remote for changing slides, microphone, or a folder with notes and logo of your institution. This object will constraint your overemphasised movements and make your hands look natural and professional.

As for the rest of your body, maintain good posture and face the audience as much as possible. This signals openness and shows that you're directing your message to them. While walking during a presentation can add energy, in more formal settings, it may be restricted to very narrow area close to the podium.

What to Do if Things Go Out of Control?

Here's an undeniable truth: something will go wrong at some point—probably more than once. It could be technical difficulty, a tongue twist, or an unexpectedly difficult question. Don't worry—these moments are a great opportunity to demonstrate your professionalism. By keeping a cool head and responding naturally, you'll show that you're in control of the situation.

Smiling during your presentation is always helpful, but a smile in a difficult moment can be even more powerful. It gives you a moment to gather your thoughts and prepares you to respond. A smile also has the power to immediately diffuse tension. Mistakes will happen, and it's important to roll with them. If a technical issue arises with one of the slides, briefly explain what was supposed to happen and then move on. You can always revisit it at the end, but don't let one mishap derail the entire presentation.

Sometimes, you may happen to do something differently than planned. You may feel like everyone noticed a mistake, but in reality, the opposite is often true. Take the FedEx logo, for example. Did you know it contains an arrow? Most people don't notice it until someone points it out. Once you see it, it's right there, between the "E" and the "x". This illustrates how our minds work—we don't notice something until we know it's supposed to be there. Similarly, your audience may not realise that something was meant to appear on slide 13, even though it's obvious to you.

If you feel the stress is starting to take over, give yourself a moment. Take a brief pause and do something that helps you relax. The best approach is to take a sip of water or glance at your notes. Whatever you do, don't rush—do it slowly. This gives you time to compose yourself and conveys confidence. Slowing down is often seen as being confident, while rushing through the problem can be seen as the sign of stress. At the end of the day this is your presentation! It moves at your paces. Own that, take a deep breath, and proceed through the slides at your own rhythm. And don't forget—feeling a bit of stress is completely normal. Everyone experiences it, and everyone makes it through unscathed. You will too.

Takeaway points:

- **Don't stare at your slides; stand facing your audience and speak to them.**
- **Be conscious of your body and use your hands with purpose.**
- **If something goes wrong, smile and take a pause.**

5.7 When to Stop Practising and Start Delivering

Two Zones

We can learn a lot from businessmen, professional musicians, and athletes. Their fame and success are rarely due to chance. More often, they are the result of hours of practice, sessions with coaches, and extraordinary perseverance. A technically flawless javelin throw or an emotionally charged guitar performance has probably been perfected through countless repetitions. The same holds true in the scientific world. If someone appears to be an effortless professional when giving a presentation, it's likely the result of years of experience and hard work.

I was wondering how to bring golf into this book, and here it is! I want to share a useful trick from Swedish golfer Annika Sörenstam. In addition to her impressive achievements on the golf course, she has also gained recognition as a coach, offering valuable advice to other professionals. One of her key pieces of guidance is to encourage players to divide their game into two spaces, separated by an imaginary line.

The Practice Zone

In the practice zone, the player focuses on honing their skills and analysing the factors that influence their performance. This is where golfers work to perfect their swing. The analysis is detailed, examining every small factor: the swing is carefully assessed with a coach, often supported by sophisticated technology, and incremental corrections are made step by step. With each attempt, the player works to control a specific element of their technique, aiming for improvement in their overall result.

You can approach your presentation preparation in much the same way, even though creating a captivating presentation might feel worlds apart from playing golf. Divide your work into two zones. In the first zone, focus on the advice from earlier sections about preparing your presentation and practising selected elements. You can follow the five steps outlined in Sect. 5.3 for preparing your presentation. This is also the stage when you should apply the insights from Sect. 5.5 to make the content more engaging and 5.6 to work on your body language.

In the practice zone, you conduct trials and critique your own work. This is also the right time to seek feedback from a colleague or video yourself. It's an opportunity to step into the mind of your audience and anticipate the questions they might ask you.

Equally important is knowing when to stop practising. At some point, you'll become fatigued, or in a sense, overtrained. Further effort may become counterproductive. When you feel truly ready and excited about the presentation, that's the ideal moment for transition out of the practice zone.

The Performance Zone

In the performance zone, the golf player should stop overthinking and focus solely on the task at hand—getting the ball into the hole in the fewest strokes possible. Interestingly, Annika Sörenstam's method is supported by research. Through MRI scans, researchers found that the best results are achieved when you compete without unnecessary thinking. So, what does this mean for those facing everyday challenges? And what does it mean for your presentation?

You simply stand in front of your audience and refrain from overthinking. In the practice phase, you focused on perfecting your presentation; now it's time to deliver. Trust your preparations and don't let that nagging voice in your head take over.

Over time, this approach becomes second nature. With each presentation, you'll feel more comfortable and confident. You will resist the urge to think, "Is my presentation going well?" This kind of thinking only muddles your mind, which Annika advises to avoid for your own benefit. Even if things don't go exactly as planned, you can always revisit them in the practice zone afterwards for your next presentation.

Takeaway points:

- **Draw an imaginary line between practicing and performing.**
- **In the practice zone, you can analyse every aspect of your presentation and train to improve selected elements.**
- **On the day of the presentation, transition to the performance zone—just deliver the presentation without overthinking.**

Chapter 6
Become Credible

> Credibility is the bridge between being known and being respected.

6.1 Why Do I Need to Promote Myself? I'm a Researcher, Not an Influencer

Exponential Growth in Research Output

Do you remember Sect. 2.8? That's the one with Ronaldo and Wayne Gretzky. I described there the concept of exponential growth of technology.

Currently, between 2 and 3 million new papers are published each year, depending on the estimation method. This number doubles every 15 years, which means that the volume of scientific publications is also growing exponentially. Several factors contribute to this trend. The primary one is the natural growth of information and knowledge. Much like technology, each new discovery paves the way for further findings and observations. Consider the research papers that have given rise to entirely new fields of study—each of these has the potential to generate hundreds or even thousands of follow-up papers. While not every research paper results in such a groundbreaking reaction, there is still a significant likelihood that a new area of research will eventually branch out into numerous subfields, much like the branches of a tree.

In addition, consider the increasing number of scientists and the immense pressure to publish. In many countries, academic careers are governed by the "publish or perish" rule, where scientific performance is primarily assessed by the number of publications. This creates a situation in which the sheer volume of publications becomes the focus of a researcher's efforts.

Don't Get Lost in the Crowd

With such a vast number of papers being published in each field, the dynamics of the publishing market are shifting. To illustrate the scale of these changes, consider

Albert Einstein's seminal paper "On the Electrodynamics of Moving Bodies". This groundbreaking work introduced the theory of special relativity, a new branch of physics, and is now regarded as one of the most significant publications in the history of science. It's accessible online if you wish to explore it further.

A contemporary reader will immediately notice a crucial detail: the paper contains no references. In today's publishing landscape, this would likely result in the paper being desk rejected. However, in 1905, when this work was published, the pace of publishing in the field of physics was so limited, that most scientists were familiar with the current state of knowledge. As a result, the absence of references was not seen as a significant issue.

Today, this would be unthinkable. Over 300 new papers are published every hour, making it impossible for anyone to stay up to date, even in a relatively narrow field. Furthermore, numerous studies indicate that a significant proportion of papers are never cited. Depending on the estimation method, around 20–30% of all papers go unnoticed by the scientific community. I refer to this phenomenon as the "black hole in scholarly publishing". Just as a black hole absorbs photons and traps them (almost) forever, papers that surpass the event horizon of visibility remain unseen by the wider scientific community.

What could explain the lack of citations? Naturally, it could be due to the low quality of the publication or an unpopular topic. However, there are instances where highly valuable papers go unnoticed for other reasons. This could include the choice of an unsuitable journal, an unappealing title, a poorly written abstract, inadequate indexing in search engines, or the wrong selection of keywords. In most cases, it's a combination of several of these factors that contributes to the lack of attention.

I wouldn't want any of your papers to end up in this "black hole". They deserve the proper recognition and readership, which is the reward for the long hours spent in the laboratory and in the front of a computer. This is precisely why this section of the book was written—to encourage you to take proactive steps in promoting your scientific work. We no longer live in 1905, times have changed. Today, ensuring the recognition and visibility of your research is just as crucial as the writing of the publication itself. And it's on you!

As Always—You Start with "Why"

Before you begin, it's worth considering why should you promote your work. As the section title suggests, you might say "Why do I need to promote myself? I'm a researcher, not an influencer!" I already mentioned the argument of a "black hole" in relation to research papers, but let's broaden the perspective and think about the possible goals of promoting your work. I've asked this question to numerous researchers during the workshops, and here are some typical responses.

Build your position in the field—You may want others to learn about your work. It's important to ensure your paper can be easily found and accessed by your potential readers. Half the job is writing the paper; the other half is delivering it to the reader.

Improve bibliometric metrics—When others discover your papers, they might cite you, which will improve your research metrics. This is particularly helpful when you're applying for new positions or research grants.

Demonstrate your expertise—At times, individuals or organizations seek experts to assist with specific R&D challenges. For instance, Company X may wish to implement a polymer coating for their product but might be unsure where to start. They look for an expert who can communicate effectively in a language they understand.

Popularise your field—Maybe you have an intrinsic drive to make your field more popular. By contributing to the popularisation of science, you not only increase societal knowledge but also build your own expert reputation among "non-experts" very effectively.

Each of these goals requires a different medium, approach, and message—topics I will explore in the upcoming sections. To make the most of the whole Chapter, I recommend reflecting on your own objectives:

- What is your aim in promoting your research? Are you interested in any of the goals mentioned above?
- How much time and effort are you willing to invest in achieving these aims?

Takeaway points:

- **The number of research papers is doubling every 15 years, it's easy to get lost in the crowd.**
- **Many researchers take an active approach and promote their research online.**
- **Before diving in—think of your motivation for promoting your research.**

6.2 Tools to Make Your Research More Visible

There Are (too) Many Options

After reading the previous section, you should have a clear understanding of your goals in promoting your research and the amount of time and effort you are prepared to invest in achieving them. The next step is to choose the right tools. There are many options and by the time this book is printed, new ones are likely to emerge! I will begin by grouping them into categories.

Academic Social Networks—The largest platforms in this category are ResearchGate and Academia.edu. These are excellent for promoting your research to other academics. With a well-maintained profile, you can increase the readership of your papers, boost citation numbers, and attract research partners. You will find a detailed discussion of these platforms in Sect. 6.3.

Professional social media—While there are smaller niche networks, LinkedIn and X are by far the most influential platforms of this type. They are highly effective for increasing visibility among professionals outside academia. Since these platforms attract a broader audience, your profile should be tailored accordingly. I explore them in detail in Sect. 6.4.

Personal profile—If you want to focus on showcasing your research activities, a simple personal webpage may suffice. This acts as a landing page highlighting your research output: papers, patents, projects and others (depending on the platform). Numerous services cater specifically to researchers, such as ORCID, Scopus Author ID, Publons, Google Scholar, and ImpactStory. Many institutions also create profiles for their researchers. Section 6.5 provides a comprehensive overview of these types of profiles.

Repositories—In Sect. 6.6, I will explore two types of repositories:

- **Research paper repositories**—These are open-access libraries where you can upload a research paper or its preprint. The most widely known repositories include ArXiv and PubMed, but many others cater to specific research disciplines. Additionally, most universities maintain their own institutional repositories.
- **General-purpose repositories**—These platforms function similarly to paper repositories but accept a broader range of materials. This includes datasets, figures, slides, videos, plots, and numerical codes. Examples of such platforms include Figshare, Slideshare, Dryad.

Of course, there's no need to restrict yourself to a single platform. In fact, having profiles on multiple platforms is often beneficial. However, it could easily become a time-intensive activity. Therefore, it's important to prioritise your platform choices based on your primary goals.

Audience and Message

With the vast reach of the internet, consider how far your content could travel: a fellow scientist, an entrepreneur, an investor in emerging technologies, an NGO operating in your field, or even a hobbyist deeply interested in your research. These individuals could be located anywhere in the world. However, not all of them will engage with the same type of content or respond to the same style of communication. This is why it's essential to define your target audience. Understanding who you want to reach will guide your choice of platform and shape your communication strategy effectively. After selecting a platform, keep in mind who will be looking at your profile and how you want to be perceived. Avoid cluttering it with irrelevant information—no fuss, no filler.

For instance, a profile on ResearchGate doesn't need to mention that you worked in industry for two years. Most users visit this platform to find scientists and their publications, so your message should be highly focused. Include your papers, grants, and research projects.

LinkedIn, however, serves a different purpose. It's not ideal for listing all your publications. Instead, you might share one or two key papers with additional explanation for non-specialists. You'll make a stronger impression by sharing content that appeals to wider professional audience, such as presentations, videos, or links to press releases written in accessible language.

Remember, your target audience and messaging are not set in stone—you can adjust them anytime, and you almost certainly will do that many times. So don't wait for the perfect moment to craft an absolutely perfect profile. Start today and refine it over time. Even very basic profile is far better than having none at all.

It's not Just About People; It's Also About Bots

We've already considered the various people who may read your online content, but there are other important readers you might not have thought about: the bots! More specifically, the content-indexing bots used by search engines and AI assistants. I don't need to emphasise the importance of appearing high in search engine results or being present in AI assistants.

However, it's also worth highlighting that these bots "learn" the connections between keywords. For example, if a bot frequently encounters content linking your name with the words "biomechanics" and "London," it will keep that association in "mind". This boosts the likelihood that your website or profile will appear when someone searches for "London biomechanics specialist." This presents a valuable opportunity to make new connections with people you may not have otherwise encountered.

Another crucial factor is that search engines also consider the popularity of your website when determining its ranking. You've probably heard of Google PageRank, which influences how a page is positioned in search results. To improve your rank, you need to drive more traffic to your site and increase the number of links pointing to it. This is why posting your content across platforms and websites can be highly beneficial.

- **Course materials**—If you're teaching a course, consider posting selected content publicly. Students will frequently download these materials, which boosts your website traffic and contributes to improving your page rank.
- **Conference presentations and posters**—After presenting, your slides and posters often languish on your hard drive. You can give them second life by uploading to a repository—this small effort can help increase your site's visibility.
- **Use university domain**—If possible, take the opportunity to feature on your university's website. Higher education institutions typically hold high value in search engine algorithms due to the volume of traffic they receive.
- **Guest posts**—Whenever possible, seek opportunities to contribute to well-established platforms such as blogs, news sites, or popular science websites. Cross-referencing with these high-traffic pages significantly boosts your page rank.

Takeaway points:

- **There are many tools to help promote your research.**
- **Before selecting your tools, it's essential to define your target audience and clarify the message you want to convey.**
- **Publishing your work online boosts your chances of connecting with valuable contacts and improves your search engine ranking.**

6.3 Academic Social Networks—ResearchGate and Academia.edu

Choice of Platform and the Main Message

ResearchGate and Academia.edu are the most popular Academic Social Networks. In addition to the standard social networking features, these platforms offer additional tools tailored for academics. These include the ability to exchange papers, export citations, share experimental data, engage in discussions related to specific research topics, and track recent papers from your connections. They also allow you to monitor citations and other bibliometric factors. Furthermore, an increasing number of academic institutions are using these platforms to post job offers.

Should you create a profile on them? That ultimately depends on your goals. These networks are particularly effective for promoting your research to academic peers. If one of your priorities is to connect with fellow researchers, I would recommend creating a profile on at least one of them. The platform choice is largely subjective, given the significant overlap in features among them. Your decision might come down to the user base of each platform. Which one is more popular in your discipline? To sense that you can look for profiles of experts in your field on both platforms.

Build the Profile

As an example, I will use ResearchGate. Creating a strong profile involves adding a photo, affiliation, introduction, listing your disciplines and skills, and finally adding papers. Let's go through these elements one by one.

Photo—Contrary to what you might think, this is often the most challenging part of the profile. After all, who enjoys looking at photos of themselves? Every shot and angle may seem awkward. However, it's important to have a good photo because first impressions matter. Your photo should look natural. For the best result, consult your institution's marketing department. They might be organizing photoshoot sessions. Alternatively, ask a colleague with a passion for photography for help. There's nearly always one in every research group!

Affiliation—ResearchGate allows you to define affiliations at two levels.

- The first level involves selecting your university and organisational unit from a list, followed by specifying your role at the institution. Be sure to briefly describe your role to highlight the focus of your work.
- The second level is choosing the laboratory you are part of. This feature enables the grouping of collaborative projects, the presentation of publications list, and the enhancement of the research group's visibility.

Introduction—This section is the first piece of information visitors see, so it's crucial to make a strong impression. Since ResearchGate is a network for scientists and researchers, you can use jargon and specialised terms relevant to your field. However, be cautious not to go too far, as this could alienate those from other disciplines. A good approach might be to list the areas you are currently or have recently worked on. Keep it concise but informative, ensuring it reflects your accomplishments and areas of specialisation.

Disciplines & Skills—This section complements your description with a list of disciplines, skills, and areas of expertise. It's essential to make thoughtful selections here, as ResearchGate's algorithms will use these keywords to direct other users with similar interests to your profile. To maximise visibility and relevance, it's recommended to choose from the established disciplines and skills provided in the list.

Further Content Updates

Once your personal information is set, you can periodically update your profile with details about your achievements. These are mainly publications, but ResearchGate also allows uploading posters, conference presentations and datasets. ResearchGate updates the list of publications on an ongoing basis from available databases and uploads shared by your co-authors. If there are any particularly important papers, be sure to include them on your profile.

The platform also allows you to share a research paper file, provided your copyright agreement permits it. If the publisher doesn't allow you to upload the manuscript, other authors can still ask you for a personal copy. Therefore, it's important to check your ResearchGate mailbox regularly. You'll likely receive requests for the full text of your recent publications from people interested in your work. They've already found your publication and want to engage with it, so sharing the paper with them should be an easy decision.

Takeaway points:

- **Academic social networks are highly effective for connecting with scientists from other institutions.**
- **It's recommended to have a profile on at least one of these platforms.**
- **Keep your profile up to date by reviewing and refreshing it periodically.**

6.4 Professional Social Networks—LinkedIn and X

Choice of Platform and the Main Message

LinkedIn is the leading platform for professionals, so there's every reason to create a profile there as well. While it may not be tailored for researchers in the same way as the platforms mentioned in the previous section, this is precisely what makes it a valuable if you want to reach further with your message. LinkedIn is ideal for networking across various sectors—academia, industry, and the public institutions. I would strongly recommend that researchers from all fields create a LinkedIn profile as a great way to broaden their professional network.

You may also consider X. The platform's format encourages shorter, more interactive and dynamic communications. In this sense, X can be a valuable tool for promoting your research findings and publications while engaging with a broader audience. I'd recommend using it if your top priority is popularising your field.

Build the Profile

Again, I will focus on one platform and use creating a LinkedIn profile as an example. It's not just about filling in every field and creating a substandard profile. You want it to stand out. A strong profile should include these key elements: a professional photo, a concise headline, and a compelling about section. It's easy enough if you take it step-by-step. Let's break down each specific category.

Photo—This follows the same principle as ResearchGate: apply the same advice, and feel free to use the same photo for both profiles.

Short headline—The headline is 120 characters and appears just below your profile picture. It's also visible in search results or contact lists. Think of it as a brief introduction, similar to how you'd introduce yourself at a conference lunch. LinkedIn will automatically suggest a headline like "Associate Professor at Lund University" or "IT Manager at IBM," which simply states your position and company. Let's be honest—is this the most compelling way to present yourself? Likely not. So, take time to craft a headline that reflects who you are, what you specialise in, and how you can help others. While it's a challenge to fit this into just 120 characters, it's achievable. Just look around at other profiles in search of inspiration.

About—This section has much higher character limit, but that doesn't mean more text is always the better. When writing your "About" section, remember that it can be viewed by both specialists and non-specialists alike, so aim for a description that is both meaningful and accessible to a wide audience. Avoid overloading this section with jargon and technical details. Instead, focus on the broader significance of your work.

Your goal here is to demonstrate your expertise and the impact of your research. Highlight the value you've created for society, whether through important findings, papers, or projects. Share specifics but keep the language simple enough for everyone to understand. If you're passionate about something, don't hesitate to show it—enthusiasm can make your profile stand out.

If you feel your career is still in its early stages and you lack substantial impact to showcase, then think again! I'm sure there is something you have contributed to group projects, voluntary work, or research efforts that achieved something noteworthy. Don't be shy about highlighting those accomplishments. If you're unsure, you can always mention your aspirations or goals and share what excites you about your field.

Education & Work experience—At the simplest level, you can list your schools, employers, positions with the dates. This will look fine, but it won't be particularly engaging or informative. LinkedIn allows you to make this section more compelling by adding descriptions for each entry. It's a good idea to write 2–3 bullet points for each role or education experience. Highlight the key challenges you faced, your main achievements, and what you gained from it in the long run. This helps provide more depth and makes your profile stand out.

Skills—Choose skills from LinkedIn's list that best reflect your expertise. It's important to do this early on, as it takes time to gather endorsements from your collaborators. The more endorsements you receive, the more credible your profile will appear. Be strategic about selecting skills that align with both your professional focus and your target audience.

Further Content Updates

Similar to ResearchGate, once your main profile information is set up, you can systematically add the following elements to enhance your LinkedIn profile: recommendations, and additional sections. Let's go through these elements again, one by one.

Recommendations—This section is especially useful if you're looking to become a freelancer or apply for jobs after your Ph.D. Recommendations should come from people who have worked with you. While it's unlikely someone will spontaneously write one for you, you can proactively ask your supervisors or collaborators for recommendations. A great time to request one is when you're closing a section in your career, such as finishing a project or transitioning to a new job.

Additional sections—LinkedIn allows you to add a variety of sections, such as publications, patents, presentations, links, multimedia, projects, and volunteering experience. While this is a great opportunity to showcase your full range of accomplishments, it's important to exercise moderation. Only add elements that are relevant to your audience. Too many sections or irrelevant content can clutter your profile and dilute its impact. Focus on quality over quantity. For example, avoid overwhelming your profile with too many scientific publications. Instead, focus on adding the most important ones, along with a brief summary highlighting the main achievements. A press release about your project might be more engaging for a wider audience. Additionally, incorporating images and videos can effectively complement the text and make your profile more dynamic.

One of the main purposes of LinkedIn is to build your professional network. After a fruitful conversation at a conference, don't hesitate to invite your new colleague to connect on LinkedIn. Adding a short, personal note to each invitation can make it more meaningful. Please remember the 48-hour rule. If you send your invitation within that time, you have higher chance of being accepted.

Building your network on LinkedIn and X takes time, but rest assured, you'll begin to see the benefits in a matter of months or even weeks.

Takeaway points:

- **LinkedIn and X are ideal for connecting with professionals across various sectors, including industry, public sector, and NGOs.**
- **Customise your headline to offer a brief but impactful description.**
- **Building a network takes time, so don't expect immediate results—this is a long-term endeavour.**

6.5 Your Profile Across Platforms—from ORCID to Personal Page

Where to Create Your Research Profile?

A researcher's profile serves as a digital business card, providing essential information about your work and achievements. It acts as the go-to place for anyone seeking details about your research. As a PhD student, you can choose from several options for hosting your profile.

- **ORCID**—After creating a profile, ORCID provides you with a unique identifier linked to your profile. Your ORCID page includes key information about your publications and grants. The list of works can be updated by you or other authorised organisations. Because ORCID is run by a non-profit, it is considered objective and universal. You are likely to need your ORCID identifier in certain situations, such as when applying for a grant.
- **ResearcherID**—A similar service offered by Clarivate Analytics, a private company that indexes papers (more details in Sect. 4.2). ResearcherID also provides a unique identifier. The database is integrated with Web of Science, which has its pros and cons. The advantage is that it gives you deeper insights into metrics related to your publications. Another advantage is that your profile is automatically created after database recognises your first papers. The disadvantage is that you have limited control over the profile, as it's more tightly controlled by Web of Science.
- **Scopus Author ID**—This service is comparable to ResearcherID but is provided by Scopus, another commercial entity indexing papers (again, see Sect. 4.2). The little differences lie in the types of metrics they offer and a range of indexed documents.
- **Google Scholar**—Did I mention Sect. 4.2? I'm explaining there the essential difference between Google Scholar functioning as a search engine, and above-mentioned databases indexing scientific journals: Scopus and Web of Science. As a search engine Google Scholar has access to much more documents. But it does not have the capacity to create your profile automatically. It needs your help in confirming affiliation and list of papers pre-selected by the search engine. It will not take longer than 15 min to do that and make your profile from non-existent to fully working!
- **ImpactStory**—A non-profit organisation that allows researchers to create profiles. What sets ImpactStory apart is that it combines a typical list of publications with online mentions and shares across blogs and social media platforms.

It's highly likely that your institution has guidelines for creating and maintaining your online profiles, and there may also be a dedicated team available to assist you. Universities and research institutions benefit from promoting your impact in much the same way you do. It's an efficient and cost-effective strategy for improving outreach, which can lead to increased citations, new project partnerships, and contracted research.

Your Personal Website

Many universities create online profiles for their staff members. University domains tend to be highly ranked by search engines due to the high traffic they attract. This makes such profiles valuable to both human visitors and search engine bots. As a Ph.D. student, you may not have the opportunity to have a personal university profile, but you can consider creating your own website.

An advantage of a personal page is its stability. When you change jobs, your university profile is likely to disappear, but your personal page remains under your control.

You might think it's too early to create a personal website, especially if you don't yet have many papers or lecture materials to share. That's understandable, but my counterargument is that the sooner you start, the better your website will become over time. You can always begin with a simple one-page template serving as your online résumé and contact form. As you accumulate more materials, updating your page will be a breeze.

Building a personal profile is not limited to researchers alone. Photographers, musicians, architects, influencers, and game developers, for example, often have engaging personal pages to showcase their work. For inspiration, consider stepping outside your discipline. One particular type of personal page worth considering is a blog, which I will explore in greater detail in Sect. 6.7.

The final question is: how do you go about creating your website? Do not worry, you can opt for a ready-made template using Wordpress, Canva or Notion. Alternatively, there are various platforms like about.me, Linktree, and onee.page, which allow you to create and host a personal page for a monthly fee.

Components of a Personal Website

Whichever platform or method you choose, here are some essential elements to include on your personal page:

- **Natural Photo**—As mentioned in previous sections, a genuine, approachable photo is key to establishing trust and connection.
- **Description of your research work**—This section, similar to "Introduction" on ResearchGate or "About" on LinkedIn, should summarise your research interests, areas of expertise, and key achievements. Ideally, keep it concise—200 words is perfect, but up to 500 words is acceptable.
- **Keywords**—Select well-researched keywords that resonate with your field and are easy to understand for a non-expert. Google Trends is a useful tool to find popular terms and compare their search volume.
- **List of publications**—Ensure your list is up-to-date, and include direct links to your papers. For paywalled papers, you can link to their landing page so readers can access them more easily. Some publishers permit you to publish your paper on your research homepage (see Sect. 6.6 for more details).

- **Projects**—Highlighting your completed or ongoing projects can attract potential collaborators or partners. You can list them in brief bullet points or provide a short description of each.
- **Memberships & awards**—This section can help build your credibility, both within your academic field and for the general public. Don't hesitate to list your achievements—anything that sets you apart.
- **Student materials**—If you engage with undergraduate or postgraduate students, this is a great way to share course materials and attract visitors. Instead of emailing large files, you can direct students to your website for easy access.
- **Links**—Connect your social media profiles and researcher accounts (e.g., LinkedIn, ResearchGate) here. Linking to other reputable websites can improve your page's search engine ranking, and offering multiple ways to contact you increases the likelihood of outreach.
- **Contact details**—Make sure your contact information is easy to find. Having an excellent profile and compelling description is useless if visitors cannot reach you.

Takeaway points:

- **You should have at least one profile dedicated to your research.**
- **Your Google Scholar profile can be created in about 15 min.**
- **Creating your own page will help you reach a broader audience.**

6.6 Research Paper Repositories

The Basics

Repository is a virtual library that allows you to upload your papers, ensuring they are properly cataloged. Each file gets its own landing page containing details such as authors, abstract, submission date, and journal information. Repositories allow users to search for papers or organise them by subject.

There are many repositories tailored to specific scientific disciplines (like ArXiv or PubMed Central), countries, or institutions. It's likely that your university has its own repository. I always encourage Ph.D. students to upload their papers to at least one repository, as it's an easy way to increase the outreach. The paper will be more visible to readers and search engines. This is particularly beneficial because respected repositories are well recognised by Google Scholar. This means that adding your paper to such a repository is akin to adding it to the most widely used search engine in the world—quite a good deal!

However, when discussing research repositories, I often get asked: "Does copyright agreement with publisher allow me to do this?" The quick answer is—it depends. This section gives a little bit longer, but also more useful answer. At this point, it's important to consider both open access and paid access publishing models.

6.6 Research Paper Repositories

- **Open access papers**—If you publish your paper using an open access (OA) model (as discussed in Sect. 3.1), the copyright is not transferred to the publisher. This means you, as the author, can use the paper in any way you see fit. The downside, however, is that OA often requires you to pay an article processing charge (APC), since the publisher cannot make money by selling your paper.
- **Papers published in the paid access model**—The situation becomes more complicated in this scenario. The advantage is that the author doesn't have to pay for publication as the publisher will earn money on selling it to readers. To do this, however, the publisher requires the transfer of copyright. As the author, you cannot distribute your paper for free without the publisher's consent. Be aware that this also applies to specific parts of your paper. For example, if you want to reuse a figure you created and published in the paper, you'll need to ask the publisher for permission.

Navigating Publisher Policies

I can imagine that you have just started anxiously searching through your drawers for copyright transfer agreements. Don't worry—journals often outline specific "paths" that allow you to share your paper, even if it has been published under a paid access model.

Publishers must balance two conflicting priorities. On one hand—the journal aims to restrict access in order to generate revenue from selling paper to readers. On the other hand—overly restricted access can reduce citations, diminishing the journal's prestige. To navigate this, publishers often permit authors to share their work under certain conditions. Common examples include:

- Allowing the author to share a draft version of the paper, accompanied by a link to the final published version. This strategy can generate interest in the paper and potentially lead to additional purchases of the final version.
- Permitting the paper to be shared after an embargo period (typically 12–24 months), during which the publisher has had sufficient time to profit from it.

To simplify navigating publisher policies without a degree in law, you can use the Open Policy Finder website. By selecting a specific journal on this platform, you gain detailed insights into the publisher's policy regarding repositories.

The process involves identifying the version of your paper you wish to share, as publishers typically define rules for three different versions:

- **Submitted version**—The draft of the paper submitted to the publisher before peer review.
- **Accepted version**—The draft revised to incorporate peer reviewers' comments but not yet formatted or typeset by the publisher.
- **Published version**—The final version released on the publisher's website, complete with branding and bibliographic details.

For each version, Open Policy Finder lists multiple "pathways" outlining specific conditions for sharing. You may need to review all these pathways to understand your options. Key elements of the pathways include:

- **OA fee**—Indicates whether you must pay an open access fee for this sharing option.
- **Prerequisites**—Specifies any conditions that must be met. For instance, research funded by specific organisations (e.g., the European Research Council) may have dedicated agreements with the publisher.
- **Embargo**—States if the paper can only be shared after an embargo period, typically 12–24 months.
- **Location**—Identifies the permitted platforms or repositories where the paper can be shared under these rules. For example, you might be allowed to present the paper only on your personal website.
- **Publisher deposit**—Indicates whether the publisher will automatically deposit the paper in a repository after the embargo period.
- **Copyright owner**—Clarifies whether the author or the publisher holds copyright for the paper.
- **Conditions**—Lists requirements, such as including a link to the publisher's version or the DOI number alongside the shared paper.
- **Notes**—Offers additional relevant details specific to described pathway.

Embargo periods can pose a challenge, as many authors forget to upload their papers to a repository once the embargo ends. However, there are ways to address this. Some journals handle the process automatically by depositing papers into repositories when the embargo expires. Alternatively, many repositories allow you to upload your paper in advance and specify the embargo end date. This ensures the paper is released automatically at the appropriate time.

Repositories Beyond Research Papers

So far, I've focused exclusively on repositories for research papers. However, as a researcher, you're likely producing a wide range of valuable content beyond papers: datasets, software, images, posters, infographics, presentations, teaching materials, and more. All of these can also be deposited into appropriate repositories.

Why should you consider the repositories beyond those intended for research papers?

To reach different audience—Not everyone is interested in reading research papers. Some people may be searching for software, datasets, or other materials relevant to their work. By depositing these resources in specialised repositories, you increase the visibility and recognition of your work among diverse groups.

To address the replication crisis—Science faces a replication crisis—many studies cannot be replicated due to insufficient or inaccessible data. A 2016 Nature survey revealed that 70% of researchers had encountered failures in reproducing results from other studies. Sharing datasets and supplementary materials is one of the most effective ways to combat this issue.

6.6 Research Paper Repositories

To enhance the impact of your papers—Many publishers now request datasets and supplementary materials during paper submission. This refers back to the replication crisis but also boosts journal metrics—additional materials make papers more appealing and frequently cited. Whether you deposit these materials in a publisher's repository, a third-party repository, or both, their availability increases your paper's impact. Some publishers even automatically share figures and datasets in external repositories.

To maximise the value of your work—You invest significant time preparing conference presentations, posters, codes, and datasets. Often, these materials sit unused after being presented. Uploading them to a repository gives them a second life, allowing others to benefit from your effort while maximising the impact of your work.

To protect your work—Reputable repositories address copyright issues and provide tools to safeguard your work:

- **Tracking**—Repositories log publication dates and all amendments, offering proof of originality and authorship.
- **DOI Assignment**—By reserving a Digital Object Identifier (DOI), your work receives a unique, permanent link, ensuring it can always be identified and cited.
- **Copyright Agreements**—Many repositories support widely recognised copyright standards, such as Creative Commons licenses. For instance, a CC BY-NC license allows others to use your work as long as they attribute the source and avoid commercial use.

Examples of repositories for content other than research papers:

- **Figshare**—A general-purpose repository initially designed for sharing academic figures and graphics. Today, it supports nearly all file formats, allowing users to upload and organise content into collections or create collaborative spaces.
- **Dryad**—A specialised repository dedicated to sharing datasets. It offers a structured approach where datasets are reviewed before publication. While Dryad charges a small fee to cover data curation and preservation costs, it remains a non-profit organisation.
- **Zenodo/Harvard Dataverse**—Zenodo, backed by CERN, allows the publication of datasets, documents, and more, while Harvard Dataverse, supported by Harvard's Institute for Quantitative Social Science, Libraries, and Information Technology, provides similar functionality. Both repositories ensure high credibility and support for sharing a variety of academic resources.
- **Software repositories**—For coders, software repositories are essential. While GitHub is the most widely known, there are also repositories for particular programming environments like MATLAB Central or PyPI.
- **Slideshare**—Originally designed for sharing presentations and documents, Slideshare has evolved through acquisitions by LinkedIn and Scribd while maintaining its core functionality. It provides an effortless way to upload and publish presentations, extending the life and reach of your work. An effective practice is

to include a QR code on the final slide of your conference presentation, linking to the uploaded version for easy audience access.

Takeaway points:

- **Placing paper in a repository significantly enhances its visibility.**
- **Always check the journal's copyright policy before uploading your paper to a repository.**
- **You can use dedicated repositories for sharing presentations, code, datasets, and posters.**

6.7 Popular Science Communication

Expanding Your Reach

Until recently, it seemed that blogs, podcasts, videos, newsletters, and social media channels were primarily the domain of businesses or influencers in fashion and lifestyle. However, in recent years, these platforms have gained significant importance among scientists.

Let's focus on blogs, though the principles discussed here can be applied to other media as well. A well-written popular science article can reach far broader audience than specialist papers published in academic journal. I know this might be a tough pill to swallow, but the good news is that there's plenty of room for everyone. The key is ensuring that your work is written in clear, accessible language and style, with a message tuned for a wide audience.

Before I dive into how to do this, let's first explore the benefits of sharing your knowledge in a popular science format—both for you as a researcher and for your audience. One crucial aspect of a scientist's career is building their reputation and personal "brand." It's up to you whether you choose to communicate exclusively with specialists in your field or open yourself up to broader circles. Building your brand through popular channels enables you to expand the range of people you can engage with.

More than just raising your profile, you can view this as an educational mission. Reaching the wider public offers an extraordinary opportunity to break down the stereotype of a scientist locked away in a dark lab, working on something that seems disconnected from the real world. Let's show people that the work of a scientist is fascinating! Let's demonstrate that our work is here to change the world!

Writing a popular article requires a departure from the formality of scientific writing. Travel back to when you were starting your university degree. You will need to write for that version of yourself. This is the biggest challenge, but also a valuable exercise for your mind. It will make you a better communicator—an essential skill in many areas. By developing it you'll improve your conference presentations, enhance your ability to engage with industry, and develop a stronger skillset for writing compelling grant proposals. So there's much more at stake than just one article.

Writing for a Wider Audience

Starting is always the hardest part, but as with anything, the more you do it, the more natural it becomes. A great way to ease into this is by making a guest post on your university's website or social media channels. If that's not an option, contributing to an external blog or podcast is a good alternative. This approach will help you get comfortable with the process, and the people involved will likely offer valuable feedback early on.

A popular science article should take around 5 min to read—roughly 1,000 to 1,500 words. While it might be tempting to dive into technical language, it's best to avoid jargon. By now, you should be comfortable simplifying your message, but this piece should be even more straightforward than your personal page or LinkedIn profile. Remember, the reader may have little to no knowledge of your topic. To explain a complex concept, try using a comparison or analogy from everyday life. For example, one of my favourites is an analogy of oarsmen in a rowboat to explain the concept of the gene survival, as explained by Richard Dawkins in his book "The Selfish Gene".

When using figures, try to relate them to dimensions that are familiar to everyone. A great example of this comes from L. Lederman and D. Teresi's book "The God Particle". They explain the size of an atom by comparing its nucleus to a ping-pong ball. If the nucleus were that size, the size of the atom would be roughly equal to the size of a baseball field. This is something people can visualise through their own experiences, making it much easier for them to grasp. The key is to put it in terms that resonate with your audience's everyday life.

Storytelling is a powerful technique that deeply engages the reader. Try starting your article with a short story that your audience can relate to. You can read more about this approach in Sect. 3.2, where I discuss storytelling in academic papers, or in Sect. 5.5, where I provide examples of using it in presentations. Yes, I repeat this advice in different contexts because it has proven effective in all of them. For popular science writing, you simply need to elevate it by finding a story that resonates with everyone.

Once you've written your article, do the "Mum test". Give it to someone you trust, but who isn't involved in science, just like your mother! Ask her how easy it is to understand, but don't worry about comments that might not seem constructive. Focus on what she understood from your text and consider whether that aligns with the message you want to convey.

What if You Love It?

If you enjoy popularising science, it might be time to consider starting your own blog, website, mailing list, podcast, YouTube channel, or even more influencer-like presence on social media of your choice: LinkedIn, Facebook, or TikTok. While this will undoubtedly be more time-consuming and require some basic marketing knowledge, it offers complete freedom in shaping your personal brand and selecting the topics you want to explore. Most importantly, it allows you to build a community of people who care about your work. Once you've built your audience, you'll be well-positioned to bring in guests and turn your channel into a productive content platform.

The key to success lies in finding your own unique online presence. Following successful channels in your space lets you observe those who've already reached your desired level. This allows you to gain inspiration and, over time, discover the niche you can fill. Eventually, you might even consider reaching out to a science populariser you admire. Who knows? It could lead to your first guest appearance—hopefully, the first of many.

I had the privilege of witnessing the rise of the "Fire Science Show" by Wojciech Węgrzyński, one of the success stories worth sharing. It's a truly fascinating podcast series. I met Wojciech a few years ago at one of my workshops, when he was just looking for ways to improve his research papers. Over time, he transitioned to the idea of creating a podcast about fire research, which has now gained worldwide recognition, a large community, and stable sponsorship.

In a recent conversation, Wojciech shared that he initially started the podcast as a fun project to satisfy his internal drive for sharing knowledge. He told me, "If someone had told me back then, how much I would gain from this as a researcher, I would have thought they were crazy." Now, he's inviting some of the most respected individuals in the field to appear on his podcast, and they consider it a great honour. He has established connections with top laboratories. Wojciech said, "This has been the most inspiring and engaging thing that could have happened to me as a researcher".

Takeaway points:

- **Why not try writing an article for a wider audience?**
- **Start small with a guest article and see how you feel about it.**
- **If you enjoy it, you can consider running your own gig, which will also help you become a better researcher.**

6.8 Stepping into the Reviewer's Shoes

Am I Ready?

Becoming a reviewer largely depends on your mindset and the time and dedication you're willing to put into the process.

First and foremost, you need to feel confident in your ability to evaluate the work of others. A solid understanding of your field is essential. Regardless of how specialised your area is, it's important to stay up to date with the current state of knowledge, recent findings, and key publications. By this stage, you should have already published your own papers.

For early-career researchers, the benefits of acting as a reviewer are particularly valuable. It offers a unique perspective on the peer-review process, providing insights into how journals and editors operate. You'll stay informed about the latest literature and, crucially, hone your critical thinking skills. In my view, reviewing the work of others ultimately helps us become better authors and researchers ourselves.

And just pause for a second to reflect in which book chapter this section is located in. It's "Become credible" for a reason. I truly believe that serving as a trustworthy reviewer is another important way of building your reputation among your peers.

Becoming a Reviewer

If you feel confident about taking on the task of reviewing someone else's work, the next step is to find a suitable journal. Look up the contact details for the editor-in-chief, as you'll need to send a personal email in order to express your interest in becoming a reviewer. In your letter to the editor, be sure to include the following:

- The institution and research unit you represent.
- The field of research you specialise in.
- Why you want to become a reviewer.
- Why you've chosen this particular journal.
- Your experience as an author, citing your published works or attaching your CV.
- Your availability to meet editorial office deadlines.

As a journal editor, I often receive emails like this, and I always respond positively. Almost every journal editor struggles with a shortage of reviewers and is eager to add new, enthusiastic individuals to their database. Those who are committed to providing quality reviews are especially valuable.

After reading your letter, the editor may wish to check your researcher profile on ORCID or another platform. This is yet another reason why it's beneficial to have your profiles set up and ready, as discussed in previous sections. These platforms provide editors with a convenient way to learn more about you and your research activities. If your profile checks out, you may be assigned your first paper to review. This might be one that has already been reviewed or even published. The editor will use this as an opportunity to evaluate the quality of your review and compare it with feedback from more experienced reviewers.

Writing a Valuable Review

Once you receive your first paper to review, it's time to get started. Based on my experience, here's what defines a good reviewer.

Be realistic—A good reviewer makes practical judgments and recognises limitations:

- **Your specialisation**—Surprisingly, many reviewers take on papers that fall far outside their expertise. Before accepting a review assignment, read the abstract to confirm that it aligns with your area of knowledge.
- **Your availability**—Reviewers often underestimate the time required for a thorough review. Remember, a well-crafted review cannot be rushed, so start early.
- **The author's capabilities**—Asking for 15 additional figures or suggesting unrelated avenues of research may be excessive. A reviewer's role is to evaluate the manuscript as presented, identifying its strengths and weaknesses. Help the author

emphasise the former and marginalize the latter. You should suggest improvements, but if substantial corrections are needed—you may need to make tough decisions. Ultimately, a reviewer aids the author but cannot overhaul their work entirely.

Be precise—The editor values clarity, so your recommendations should be specific and well-structured. Here's a list of what should be included in the review:

- **Indicate your overall recommendation**—Clearly state the overall recommendation you suggest: accept, minor revision, major revision, or reject.
- **Pros/cons**—As an introduction, highlight the work's strengths and weaknesses in a concise format, e.g. short paragraph. This summary helps both the editor, and authors understand the key factors behind your overall recommendation.
- **Detailed list of comments**—Move on to providing page-by-page feedback with specific comments. Use a structured format, such as numbered list or a table, to ensure clarity and traceability. For each comment, include:
  o **Location**—Specify the page, section, or sentence you are refering to.
  o **Issue**—Describe the problem. What is the issue, and why is this important?
  o **Expectation**—Make your comment clear. Writing "this section needs improvement" might not be well understood by the authors. If you spend a little more time to explain precisely what sort of improvement you are expecting, the chances of getting acceptable revision are much higher.
- **Message for the editor**—In many cases the journal system leaves the opportunity to send note directly to the editor. It is a great chance to thank for the opportunity and share key recommendations regarding the decision process. You can indicate whether you're willing to review the revised manuscript, or suggest that minor comments could be handled by the editor directly. This approach helps streamline the process and saves time for everyone involved.

Be conscientious—A significant drawback of the peer-review process is the inefficiency of reviewers in identifying errors in manuscripts, as demonstrated by Richard Smith's widely discussed experiment (search: Slay peer review 'sacred cow'). Smith submitted a paper containing eight major errors (each sufficient for outright rejection) to 300 reviewers. The best reviewer identified five errors, while the median was two. Most alarmingly, 20% reviewers din't find any mistake. These results underscore the critical need for professionalism and thoroughness in reviewing. A good review isn't completed overnight; it requires 2–3 focused sessions. My typical process involves an initial session for reading the paper and annotating it with general notes, followed by a break to gain perspective. I then use 1–2 more sessions to consolidate my comments into a Word document and finally paste them into the submission system. This structured approach ensures accuracy and depth in the review.

Be ethical—The peer-review process has always relied on principles of integrity and fairness, and it's our duty as reviewers to uphold these standards. Avoid any unethical behavior such as:

- Reviewing papers authored by colleagues or friends with whom you have a personal or professional relationship.
- Using your role to encourage authors to cite your own work.
- Reviewing papers on topics closely aligned with your current research, particularly if you perceive the authors as competitors. Early access to their results in this context would be unethical.
- Sharing your review or its content with others, as this compromises the confidentiality of the review process and exposes sensitive information before it becomes public.

Takeaway points:

- **Reviewing papers sharpens your skills as an author.**
- **Don't wait for an invitation—identify a suitable journal and volunteer as a reviewer.**
- **Allocate enough time for the review process and be realistic, precise, conscientious, and ethical.**

6.9 Should I Consider Research Mobility?

Benefits of Pursuing Mobility

Building credibility extends beyond the virtual world. Sometimes, it requires making tangible moves in the physical space. This is where research mobility comes into play. Let's explore its advantages.

Visiting another institution can be a daunting task. It requires stepping out of your comfort zone, navigating administrative challenges at both institutions, and reorganising your personal life. In short, it's no small feat. Despite these hurdles, many students, lecturers, and professionals choose to temporarily move abroad for their work. What drives them to make this leap, and is it worth it? Without a doubt, yes! Speaking from my own experience, having worked in different countries, such opportunities offer a unique chance to embrace a new way of life, providing invaluable benefits not only to your professional career but also to your personal growth.

Professionally, a change of working environment can be incredibly revitalising. That persistent problem you've been grappling with, might suddenly seem more manageable when viewed through a fresh lens and with input from a new group of people. At the new institution, you may encounter methods and approaches that previously seemed just theoretical. You'll gain insights into different ways of organising work, conducting research, and exchanging ideas. This experience has the potential to leave a lasting impact on your entire career.

Don't Forget About the Ph.D.!

In long run—mobility will make you a better researcher. It's difficult to come up with innovative ideas when every day presents the same views outside your window, the same people around you, the same desk, the same cultural context, and the same laboratory. It takes a true genius to innovate under such repetitive conditions. I don't consider myself a genius, so I make it a point to explore unfamiliar and exciting paths. New experiences can ignite creativity, potentially leading to a personal golden age of research.

When planning a visit that will invigorate your academic career, it's crucial not to lose sight of your current priorities. Relocating shouldn't cause your dissertation to fall behind. Ideally, the trip should align with your project or be closely related to the problem you're tackling. If it doesn't directly connect with your Ph.D., it's worth discussing the potential impact on your work with your supervisor. Together, you can assess whether the benefits of the visit outweigh the challenges posed by your absence.

If your Ph.D. timeline is at risk, your supervisor may not be supportive of the internship idea. In this case, it might be worth considering pursuing an interesting post-doctoral position after finishing your Ph.D. This can also be beneficial for you—it's better to complete your Ph.D. and embark on a new adventure with a clean slate, rather than jeopardising your Ph.D. for the sake of immediate opportunities.

The Logistics

Planning a research visit, no matter the duration, is a major logistical undertaking. You'll need to identify the host institution, establish the necessary connections, secure funding, and determine the optimal timing. Given these demands, it's essential to start planning well in advance.

Where to go—Do you have a list of research groups that align with your Ph.D. research? Here are some potential ways to connect with them:

- **Your supervisor's or coworkers' contacts**—People you know can offer introductions to institutions or individuals in your field. A direct and personal introduction is often the most effective way to get a positive response.
- **Social media channels**—If you've built a network over time, platforms like LinkedIn can help you reach out to potential mentors or institutions.
- **Conferences**—Conferences are a great opportunity to connect directly with attendees from institutions you're interested in. In-person interactions will have a big advantage here—Sect. 5.2 will help you with this.
- **Cold emails**—While less effective, reaching out to people you do not personally know can still work. Tailor each message to explain why the institution is a good fit for your research and how you can contribute to their work. The more specific and personal the email is, the higher the chance of receiving positive response.

When should you go—In your Ph.D. timeline, identify a specific period when you do not need access to specific equipment available at your home institution. This could be during a time dedicated to data analysis (where an external mentor might offer useful insights), or when conducting surveys. Planning this in advance is key to aligning your tasks and ensuring that your thesis submission isn't delayed. If you do it wisely, the internship can even speed up or improve your work.

6.9 Should I Consider Research Mobility?

How to fund your visit—Let's talk about money! The hosting institution is unlikely to fund your project. Therefore, you need to look for funding opportunities well in advance. Consider the following options:

- Your university may have a dedicated early career researcher mobility program.
- National grant providers often support exchange programs.
- Embassies may have information on mutual agreements for scientific exchanges, e.g. Fulbright or Erasmus + .

Each funding source has its own rules and conditions, so checking them early will help you avoid disappointment or missing deadlines.

Final preparations—Assuming everything has gone smoothly, you'll be preparing for your visit. Organising the logistics of the trip – no matter how long it lasts—is a valuable lesson in patience. Many details need to be sorted out, with accommodation being the most obvious. It's always worth checking if the hosting institution offers any budget housing options.

Takeaway points:

- **A research visit boosts your chances of building a successful academic career.**
- **Ideally, the visit should align with your Ph.D. research topic.**
- **Early planning is essential to secure the necessary resources and an exciting host institution.**

Chapter 7
Develop Personal Toolkit for Success

> Success favours those who sharpen their tools before the challenge arises.

7.1 Your Morning Routine Is Your Friend

Consistency and Effectiveness

Whether you're an early bird or a night owl, we all start the day at some point. And it seems like we all start it differently. Some of us jump online to check out social media, others dive into emails. Some drink kale smoothies, others are rushing to pack their kids a lunch box. There are a million different ways your morning can go.

While there's probably no perfect morning routine that fits everyone, we can learn a lot from the morning habits of successful people, as well as from studies that confirm the positive impact of a morning routine on the rest of the day.

A morning routine is a set of activities we do in the morning, usually before the start of the main activity of the day, such as going to work or school. Depending on your schedule, it can take anything from a few minutes to a couple of hours.

Discipline Equals Freedom

Consistently sticking to the same morning activities generates useful habits. A solid morning routine sets us up for a productive day, leading to better results. When your morning goes as planned, you start the day with a sense of accomplishment, regardless of any challenges that arise later. Each small victory boosts your motivation, making it easier to achieve your goals. Win the morning, and you significantly increase your chances of winning the entire day.

I like the saying of former Navy Seal Jocko Willink: "*Discipline equals freedom*". This concept applies seamlessly to how you start your day. Morning discipline involves following a predetermined sequence of actions without hesitation or internal debate. You simply execute the routine on autopilot. Over time, these actions solidify into habits, allowing your brain to handle them automatically. For instance, do you consciously think about how to brush your teeth? Of course not—your habit takes care of it effortlessly.

Your Morning Routine

Your morning routine is an ideal opportunity to fit in activities you might struggle to prioritise later in the day. While it's important to craft a routine that suits your specific needs, exploring the routines of others can provide valuable inspiration and ideas. Consider including the following:

- Exercise.
- Meditation.
- Yoga.
- Reading.
- Writing, such as in a journal.
- Taking a cold shower.
- Walking with the dog.
- Brainstorming an important idea.
- Taking a small step towards a larger project.
- Spending time outdoors.
- Hugging a loved one.

My morning routine is continually evolving, but I've recently settled into a rhythm that feels productive and balanced:

- Drinking a glass of water with added minerals.
- Walking the dogs with gratitude and excitement for the day ahead.
- Practising a language using the phone app.
- Reading for 10–15 min.
- Reviewing the day's schedule, which I prepared the evening before.
- If I am commuting, listening to a podcast.

How long should you stick to a morning routine before changing it? As long as it works! I personally review mine every month—not because it's ineffective, but to see if there's a better way. Experimentation is key to finding what works best.

Takeaway points:

- **A morning routine is your ally—it provides consistency and effectiveness.**
- **Keep it simple, it should run on autopilot.**
- **Begin your day with a few small wins that set the tone for the rest of it.**

7.2 The Only High-Performance Supplement You Need

Why We Sleep

As humans, we're driven by the desire to constantly improve. Look around, and you'll see this everywhere, including in academia. We're all racing—sometimes literally—to find ways to become more productive, write better papers and make better experiments. Each of us has our own method to stay motivated, whether it's through dietary supplements, coffee, or exercise. However, there is one universal strategy that can make everyone more effective on a daily basis. It's also regardes as quite plesant! What is it? Simply—a good night's sleep.

The Sleep Foundation highlights that humans are the only mammals who intentionally cut back on sleep. In our busy lives, we often view sleep as a luxury, with more rest sometimes seen as laziness. However, this mindset is detrimental. So, why is sleep so vital? First and foremost, adequate sleep ensures our immune system functions optimally, helps maintain healthy blood sugar levels, and can even lower the risk of Alzheimer's disease. For most adults, the ideal amount of sleep for good health is 7–8 h each night. Anything less, and you'd be considered sleep-deprived by your doctor.

Moreover, regular, adequate sleep makes it easier to control your appetite. Sleep deprivation increases the levels of the hormone that triggers hunger, which is why losing weight becomes so challenging when you're not getting enough rest. While these are long-term effects, sleep also has immediate impacts. Can you be certain your brain is functioning at its best when you haven't had enough sleep? Will that new paper be your best work?

Our ability to remember and learn is significantly diminished when we're sleep-deprived. Beyond that, we become more anxious and distracted. To summarise, not only do we learn more slowly (if at all), but we also make each subsequent day progressively harder. In practical terms, this means we risk producing sloppy work and could be more prone to accidents, like a car crash, or strain personal relationships due to a short temper. Numerous studies suggest that sleep deprivation is linked to poorer mental health and an increase in symptoms of anxiety and depression.

The title of this section is no accident. If you'd like to explore this important topic further, I highly recommend the book "Why We Sleep" by Matthew Walker which summarizes a lot of scientific data on sleep.

Your Sleep Toolkit

I hope I've convinced you to prioritise getting eight hours of sleep each night. But what if you're unsure how to sleep well? Don't worry—you're not alone. Around 50% of the population struggles with insomnia, with men more likely to suffer from it than women. To boost your productivity and improve the quality of your sleep, try implementing these adjustments:

- **Room Temperature**—Keep your bedroom temperature as close to 18.3°C (65°F) as possible, as this is the value recommended by the Sleep Foundation.
- **Blue light**—Avoid using your phone, computer, or other devices that emit stimulating blue light at least one to two hours before your bedtime.
- **Stimulants**—Limit stimulants before bedtime. Caffeine, nicotine, and even alcohol (contrary to popular belief) will make your sleep less restorative.
- **Eating**—Avoid heavy, hard-to-digest meals before bed. Try to have your last meal at least two to three hours before going to sleep.

- **Winding down**—Have a calming pre-sleep routine, such as meditation, reading, or other relaxing activities, to help your body shift into rest mode.
- **Routine**—Maintain a consistent bedtime, even on weekends. Delaying your bedtime by just an hour is like switching to a different time zone, disrupting your body's natural rhythm.

You can use technology to track and enhance your sleep quality. Devices like the Oura Ring offer detailed insights into your sleep patterns, helping you develop healthier sleep habits. I personally found it extremely useful for gaining a better understanding of my sleep cycles and making adjustments where needed. However, there are many other alternatives available, such as the Whoop Strap, Fitbit, and various sleep tracking apps. Each of these options provides unique features, from monitoring heart rate variability to tracking sleep stages, all of which can support you in creating a more consistent and restful sleep routine.

Finally, here's a simple sleep recipe recommended by Ben Greenfield, an ex-Ironman triathlete, Spartan racer, coach, speaker, and author:

"Before bed, avoid caffeine for 10 h, food for 3 h, work for 2 h, and screens for 1 h. Then, in the morning, resist hitting the snooze button!"

Takeaway points:

- **Good sleep enhances both mental and physical performance.**
- **The quality of your work will decline even after one night of poor sleep.**
- **Establishing a consistent sleep routine helps your body rest effectively.**

7.3 Habits, Habits, Habits

You Are in Control

Each person is shaped by a set of habits. Some of these are positive—like regularly drinking water, doing relaxing exercises, or saving money. Others are less helpful—such as procrastination, smoking, or evening snacking. The good news is that we have the power to shape our habits, eliminating the harmful ones and replacing them with those that enhance our quality of life.

It Will Require Effort

Christine Whelan, a sociologist from the University of Wisconsin-Madison, studies happiness, human ecology, and habits. When asked by the Washington Post how long it takes to make a lasting change, Whelan estimated it takes around 60-90 days to turn a new action into habit and part of our lifestyle. Other research suggests that only 40% of our daily activities are conscious decisions, while the rest we perform almost automatically.

Charles Duhigg, in "The Power of Habit", presented a visual way of describing this automatism as the "habit loop". Each loop begins with a cue, which subconsciously triggers the loop. Once activated, the mind initiates an action without any deliberation. The loop concludes with a reward—a positive feeling that reinforces the behaviour.

A simple example is the feeling of tiredness (cue), which prompts someone to turn on the coffee machine (action) and enjoy a caffeine boost from a warm beverage (reward). By observing yourself, you can identify the cues that drive specific actions and the rewards you gain. You can then modify the loop by replacing some of its elements. For instance, instead of coffee, you could opt for other alternatives that deliver enjoyable rewards. Taking a short walk followed by a glass of water might be worth trying. You can also steer clear of triggers that can lead to unwanted cycles. Are you often tempted by a donut in the campus cafeteria? When you go to that area, take an apple with you.

The uncomfortable truth about changing habits is that, like any change, it requires effort and a realistic self-assessment. Often, someone inspired to change will create an ambitious list of new habits, many of which may be unrealistic to implement immediately—or at all. Take, for example, the goal of immediate change form not exercising to visiting the gym three times a week for 1.5 h. Achieving this would mean giving up another activity. What happens to those old activities? Do you stop doing them or reschedule? Will childcare need to be arranged, or will you need to fit it around your work schedule? It's a complex task. If this is one of many radical, time-consuming changes you are planning with the 1st of January, the questions quickly multiply. Over time, it's easy to become overwhelmed and give up. So, how do you introduce new habits wisely and effectively?

Your Habits

There are numerous studies on how to effectively implement change, but most are based on several universal principles.

Name your goal—Avoid vague aspirations like "I want to be happy". Instead, focus on specific, actionable steps that will lead you to greater happiness. For example, you might aim to spend more time with friends or get outdoors more often.

Make your goal measurable—To track progress, ensure your goal can be quantified. For instance, if you want to drink more water, keep a daily record of how much you drink. This allows you to evaluate your progress and adjust accordingly. Remember to confront your expectations with reality—some goals may not be immediately achievable.

Choose a goal aligned with your values—Implementing a change is easier if it also aligns with your deeper values. For example, if you want to eat less meat for health reasons, but also because you believe in reducing its environmental impact, the goal becomes more meaningful and motivating.

Reward yourself for progress—Celebrate your achievements along the way. If you're making progress, treat yourself to something enjoyable, like Belgian chocolate or a coconut latte.

Find support in others—Surrounding yourself with like-minded individuals can greatly increase your chances of success. But what if you can't find anyone in your immediate circle? Consider seeking support through online forums, personal trainers, or apps. Apps like Way of Life, Water Tracker, Habitify, and Stikk (which helps you stick to your commitments) are particularly helpful in reinforcing new habits.

No matter what habit you choose to implement, recognise the power you hold. Understanding that you can create our own habits will give you a powerful tool to shape your life. By making a change today, you can set in motion incremental improvements that will lead to positive, long-term transformation.

Takeaway points:

- **Aristotle was right: we are what we repeatedly do. Excellence is not a single act, but a habit.**
- **You have the power to control your habits.**
- **Recognise and eliminate bad habits, replacing them with positive ones.**

7.4 Small Wins, Big Impact

Persistence and Compounding

The power of compounding lies in its ability to generate substantial growth from consistent, small efforts. Just as compound interest in investing turns regular contributions into substantial wealth over time, your small daily victories can accumulate to create significant progress during a Ph.D. journey. Each completed task and incremental improvement builds upon the last, producing a multiplying effect that amplifies your overall impact. Over time, this compounding effect transforms seemingly minor actions into meaningful achievements.

As Naval Ravikant, the Indian-born American entrepreneur and investor said: "*Play long-term games with long-term people. The power of compounding in knowledge, relationships, and money is astonishing over time.*" And there is no reason why you shouldn't use the principles of compounding in your Ph.D. work. Whether it's writing 500 words a day, running a single experiment, or analysing one dataset, each small win contributes to steady growth towards your goal. Over time, these efforts build momentum, and what once felt unattainable becomes inevitable through the persistent application of the compounding effect.

Process Versus Result

Almost every Ph.D. student I've known—myself included—has gone through long periods where they felt as though they weren't making any progress. It can feel like you're failing, and you may even want to throw in the towel. If you've ever felt that way, you're not alone.

The sheer magnitude of completing a Ph.D. can be overwhelming. It's a daunting task, and it can paralyse you into inaction. However, small, incremental improvements add up to significant progress over time. By focusing on daily growth, even in tiny steps, you can make massive strides in the long run.

Seth Godin puts it well: "*The thing is, incremental daily progress (negative or positive) is what actually causes transformation. A figurative drip, drip, drip. Showing up, *every single day, gaining in strength, organising for the long haul, building connection, laying track—this subtle but difficult work is how culture changes."*

This is why I recommend focusing on the process rather than the result. Results don't materialise without investing time and effort into the process. This is especially crucial in the demanding, time-bound world of a Ph.D. Let's look at an example: Your goal (the result) is to submit a paper based on your research. Dreaming about the finished manuscript won't help unless you do the hard work to get there. Instead, focus on the process. Commit to spending two solid hours a day writing your paper, five days a week—no excuses. That's a clear, focused plan centred around the process.

Five Tasks Is All It Takes

There's one technique that's been crucial in driving my daily progress, and I want to share it with you. It's called the "Power List of Five Tasks", introduced by entrepreneur Andy Frisella. This method has worked wonders for me, and I'm confident it can help you too.

Andy suggests treating every day as a game. A day is won if you complete all five tasks; a week is won with at least six days of work; and a month is won with at least three weeks of consistent effort. By completing five tasks a day, you'll be certain that you're on track to achieve your goals. These tasks don't need to be performed in any specific order, and the best way to manage them is to plan your five tasks the evening before.

After several weeks of consistently winning, you'll be amazed by how much you've achieved. The tasks don't need to be enormous projects that take up most of the day. They just need to challenge you to move forward. With the right focus, you should be able to complete all five tasks by noon.

For this method to be effective, it's crucial to be honest with yourself and commit to it every day. This includes acknowledging those days when motivation may be lacking. If you miss a day, don't be discouraged. Tomorrow is another opportunity to win.

Takeaway points:

- **The earlier you begin compounding your efforts, the greater the long-term benefits.**
- **Focus on the process, and the results will follow.**
- **Set five daily tasks the evening before to ensure you're investing your energy well.**

7.5 Mindfulness

A Gym for Your Brain

We're constantly reminded to care for our bodies through regular exercise and a balanced diet. Yet few of us think about giving the same attention to the most important organ in our body—the brain. Despite making up only 2% of our body weight, the brain consumes a remarkable 20% of our energy and oxygen. It certainly deserves moments of respite from the constant influx of stimuli. But how can we effectively give it the rest it needs?

The answer lies in mindfulness. Mindfulness is not tied to any particular trend, religion, or philosophy. It simply involves the practice of paying conscious, nonjudgmental attention, commonly known as secular meditation. The concept and method of mindfulness-based stress reduction (MBSR) were developed in 1979 by Dr. Jon Kabat-Zinn, founder of the Stress Reduction Clinic at the University of Massachusetts. While this method may seem relatively new, it has proven to be incredibly effective in reducing stress and enhancing overall well-being.

There's an App for That

According to scientists from the University of Rochester, people who practice mindfulness are often characterised by their excellent mood and overall health. Mindfulness exercises for the mind help reduce stress, enhance concentration, and positively impact your quality of life. In addition, mindfulness can improve attention span, conserve energy, strengthen mental resilience, foster self-awareness, and build emotional closeness. It can also help you confront and overcome fears and even support your spiritual growth.

These benefits aren't just subjective opinions but are supported by research, including studies led by Judson E. Brewer from the Yale University School of Medicine in New Haven. Brewer and his team used MRI scans to study the effects of prolonged meditation on the brain. Their findings revealed that people practicing various forms of meditation had significantly less activity in the posterior cingulate cortex and the medial prefrontal cortex. In simpler terms, those who practised mindfulness techniques showed a reduced tendency for their minds to wander or become distracted—an outcome that scientists associate with maintaining a better mood.

If you're interested in mindfulness but aren't sure where to begin, "How to Train a Wild Elephant and Other Adventures in Attentiveness Practice" by Jan Chozen Bays is a great resource. The book offers practical ideas for incorporating meditation into your daily routine and is available in both paperback and digital formats. For maximum convenience, you can download one of several mindfulness apps, such as The Way, Oak, Calm, or Headspace, or even find guided meditation videos on YouTube. A daily five-minute session can have a significant positive impact on your quality of life.

Give It a Go

If you can't imagine sitting still and listening to a guided meditation, don't worry—start with small steps instead:

- Take a few calming deep breaths. Breathing exercises can reduce stress and ease anxiety.
- When you eat, focus on eating—avoid multitasking. Turn off the TV and put away your phone. Pay attention to the colours, shapes, and textures of the food, and really be present in the moment.

- At some point during the day, jot down a list of things, events, or people you're grateful for. At the end of the week, look back on the positive things that have happened over the past seven days.

Given the positive changes mindfulness can bring to your life, it's no surprise that people from all paths of life—whether professionals, athletes, actors, or musicians—have embraced it. Notable figures like Al Gore, Deepak Chopra, LeBron James, and Lady Gaga have all openly shared that daily meditation is part of their routine. There's a mindfulness technique for everyone, so don't hesitate to explore how this practice might improve your life.

Takeaway points:

- **You are what you think, so why not care for your most important tool: your brain?**
- **Mindfulness doesn't have to be mystical; it can be simple, practical, and grounded in everyday life.**
- **Start small with just five minutes a day—conscious breathing is a great way to begin.**

7.6 Eat the Frog First

I'll Do It Later

Procrastination is the act of intentionally delaying something, even though you know it might cause problems later. We all do it from time to time, and let's be honest, you're probably experiencing it too. It's human nature—we fall into this trap repeatedly. At its worst, procrastination can result in missing deadlines, putting you in a tough spot.

It's estimated that procrastination is a common issue for about 20% of the general population, with the rate climbing to 50% among students. The good news is that procrastination is a habit, and like any other habit loop, it can be changed or broken once you recognise the pattern in your behaviour.

Three Dimensions

So, where does the urge to procrastinate come from? There are three main factors driving it: individual character, external influences, and the nature of the task itself. Identifying and understanding what causes procrastination is the first step toward eliminating it.

The first factor is individual character and personality traits. People who are naturally more task-oriented tend to procrastinate less than those who are visionaries. Similarly, individuals with a strong sense of self-belief are less prone to procrastination, as they trust in their ability to complete tasks. Overcoming your natural tendencies is a long-term effort, but it's possible. You can break the cycle by reading relevant books, working with a therapist, or surrounding yourself with individuals who set a positive example.

Another cause of procrastination are external distractions that quietly steal our attention. We're constantly bombarded with stimuli from the world around us—gadgets like TVs, phones, the internet, and video games which are available 24/7, making it easy to get sidetracked. Thankfully, this is one of the easier aspects to manage. By ensuring your workspace is free from distractions, you can create an environment supporting focus. Additionally, you can use tools like Stay Focused, Self Control, Rescue Time, or browser extensions and profiles to block distracting sites.

Sometimes, procrastination is driven by the nature of the task itself. Tasks that are dull, tedious, or uncomfortable tend to make us delay getting started. Unfortunately, these kinds of tasks are inevitable in life. In moments like these, try to motivate yourself with external rewards. If you have a sweet tooth, a piece of chocolate could be a motivating incentive. Or, for every hour of focused work, reward yourself with 15 minutes of free internet time. These small rewards can make tackling the task at hand feel a bit easier.

Your Daily Frog

When it comes to overcoming procrastination, Brian Tracy's book "Eat That Frog!" is essential reading. It offers practical exercises to eliminate procrastination from your life in a highly accessible way. While some of his advice might seem deceptively simple, it can make a huge difference in how you approach tasks.

One of his key tips is crucial: when faced with a series of challenging tasks, tackle the hardest one first—the "ugliest" task that will cause you the most trouble. It's important to take action instead of overthinking, as thinking often leads to paralysis. Tim Ferriss presents a similar concept with his idea of "lead dominos". These are the tasks that, once completed, will make everything else significantly easier. By prioritising these tasks, you set yourself up for smoother progress in the long run.

Brian Tracy also emphasises the importance of planning before taking action. To succeed in even the smallest task, you need to be crystal clear about what you want to achieve. Write down the goal on a piece of paper, set a realistic deadline for completion, break it down into manageable stages, and create an action plan. Then, begin implementing this plan immediately. It's crucial to do something every day, no matter how small, that brings you closer to your goal. By consistently taking these steps, you build momentum and make steady progress.

The guilt, lowered self-esteem, and stress coming from missing deadlines can have a profound negative effect on your quality of life. There are many practical tools to help us take more effective action—why not use them? By managing procrastination, you're not just becoming more efficient—you're enhancing your overall well-being and reducing unnecessary stress.

Takeaway points:

- **Procrastination is a natural tendency; we subconsciously avoid difficult or unpleasant tasks.**
- **Recognise distractors in your environment and remove them when it's time to focus.**

- **Begin your day by completing a challenging task to set a positive tone for the rest of the day.**

7.7 The Pomodoro Technique

Productivity and Efficiency

Nothing boosts productivity more than focused attention on the task at hand. In the previous section, you explored procrastination. If you're aiming to boost your efficiency even further and haven't yet discovered the Pomodoro Technique, today's your lucky day. This simple yet highly effective time-management method can significantly enhance productivity.

All you need is a timer. Yes, it's that simple. It can be a physical timer, a stopwatch (preferably not on your phone), or one of the online apps, e.g. Pomofocus.

Developed by entrepreneur and author Francesco Cirillo in the 1990s, the Pomodoro Technique gets its name from the Italian word for "tomato". This name is inspired by the tomato-shaped clock Cirillo used to track his work. The concept is straightforward: break your work into 25-min intervals, known as "pomodoros", with short 5-min breaks in between. After completing four pomodoros, take a longer break of 15–20 min.

Respect Your Pomodoros

The core commitment of the Pomodoro Technique is to honour each individual pomodoro. This block of time should remain undivided, uninterrupted, and fixed. By doing this, you can give your full attention to the task at hand. Once the 25 min are up, you simply get up from your desk and take a break. If you're in the middle of a task or thought, you might worry about losing the idea during the break. However, in nearly all cases, the idea will still be there waiting for you, often with improved clarity, after your 5-min refreshment.

However, as with most performance-improvement methods, putting theory into practice is often harder than it seems. The success of the Pomodoro Technique hinges on your self-discipline and respect for your time. Maintaining full concentration for 25 min may be challenging at first. Remember, every quick glance at your inbox, answering a phone call, or reading a text message disrupts your focus and, as a result, breaks your pomodoro. If you do break the pomodoro, theoretically, you should restart it from the beginning. But you have the flexibility to decide what works best for you, as the ultimate goal is to boost your productivity—not follow rigid rules.

Get to Work

To prepare for the Pomodoro Technique, start by eliminating any distractions—put your phone aside, mute email notifications, and focus solely on your task. Choose one task to work on, preferably one that you've been putting off. Commit yourself to that single task during the pomodoro. Remember, it's only 25 min—short enough not to feel overwhelmed, but long enough to make meaningful progress. When the timer goes off, take a 5-min break. During this time take a few deep breaths, grab a coffee, stretch, play with your pet—do whatever helps you recharge. Once the break is over, dive back into the next pomodoro.

Once you have done your first pomodoros, I have some more tips for you: If you have a meeting, phone call, or any other commitment planned, it's best to delay the next pomodoro until next free 30 min in your calendar. It's not worth starting if you know the process will be interrupted. Your work environment is also crucial. If you're tackling a complex task, it might be better to avoid doing pomodoro in a busy office where the distractions from colleagues will most likely break your focus. Lastly, don't overdo it. Two sessions, each consisting of 4 pomodoros, can be quite tiring. You'll need part of the day for less demanding tasks, such as emails, meetings, or lab work.

It's a good practice to reserve dedicated time for your pomodoro sessions in your calendar. Just as you would do when scheduling a meeting, you check your calendar to find an appropriate slot and ensure there's a suitable environment for the meeting. Why not apply the same level of consideration to the most important aspect of your Ph.D.—focused work?

Over time, the Pomodoro Technique helps train your mind to enter a state of deep concentration. You'll likely find greater satisfaction in your work and progress. As you continue, your performance level will increase, and you'll notice that you're completing tasks at a pace you've never experienced before.

Takeaway points:

- **The Pomodoro Technique is effective, and free. Take advantage of it.**
- **Don't overdo it. Aim for a maximum of two sessions of four pomodoros each.**
- **Guard your daily pomodoros fiercely. This is your time to create outstanding work.**

7.8 Do You Work or Do You Create?

Two States of Mind

Each day presents its own set of challenges—whether it's a punctured tire, overdue bills, or stepping in at the last minute to cover a colleague's teaching session. These moments, where all you can do is sigh and power through, are part of life. No matter how well-organised you are, everyone has their own "fires" to put out. In such times, it's easy to get frustrated, lose patience, and even easier to forget about your long-term plans and goals.

This is in stark contrast to when we enter a creative flow. In those moments, all the problems fade away. We feel fully in control, completely engaged in the task at hand. It's as if we're guiding the direction of the wind, rather than being tossed around by it.

To tap into your full potential, it's crucial to understand and master these two states of mind: working and creating. Work helps you tackle smaller tasks and check things off your list, but progress is often slow. Creation, on the other hand, allows you to make significant strides, setting new directions in your work and life.

The Creator's Mind

The act of creation requires intense concentration, and most of us can dedicate only a few hours a day to truly creative work. However, this is enough if the time is fully committed to deep work. During this period, ideas that have been simmering in our minds finally take physical form—whether in a new design, text, drawing, or piece of code.

Creation requires a shift from a shallow, reactive state to a deep, focused mode. In this state, our mind strives to craft something that solves a problem we've been grappling with for a while. Subconsciously, we seek an elegant and potentially beautiful solution, and our mind won't settle for something merely "good enough." This is about valuing quality over quantity, as Cal Newport explores in his book "Deep Work: Rules for Focused Success in a Distracted World".

The psychological difference between working and creating is profound. Even producing something small makes us realise we've made significant progress. Sometimes, a single idea, a rough draft, or a preliminary plan can evolve into something much larger. It's crucial to give yourself the opportunity to enter this state of creation and to approach it systematically.

Your Creative Routine

Let's take a moment to consider the process of creating a research paper. Writing itself demands both concentration and creativity. You must carefully review the existing literature, describe the methods and results with precision, conduct an in-depth analysis, and propose meaningful conclusions. However, tasks such as data processing, preparing figures and tables, or formatting references do not require that intensity of creative thought.

Entering a creative state requires conscious preparation. We all have "prime time" when our cognitive performance is at its peak. For many, this is in the morning, before meetings and other responsibilities overwhelm us. Others may find their productivity peaks in the evening, once the chaos of the day has subsided and they can begin to unwind.

When we're interrupted during the creative process, our flow is lost. It takes time and extra effort to return to that state, let alone resume what we were doing. This is why it's best to allocate at least one to two hours block of undisturbed work on each day where we want to create something meaningful.

That being said, shallow work is not inherently inferior. It's still an essential part of our day, but it can easily dominate our time. Everyone can spend all day in mailbox. The key is to balance both your "creative hat" and your "working hat" throughout the day. In my experience, maintaining an 80:20 ratio (working:creating) is the minimum needed to make significant progress on your creative projects. To ensure this happens, block out sufficient time in your calendar for creative work. If you don't do that, others will fill that time with their requests and priorities.

Takeaway points:

- **Work leads to incremental progress, while creation sparks big leaps forward.**
- **Allowing your mind to create can lead to surprising and impactful outcomes.**
- **Make time for both working and creating in your day.**

7.9 Choosing What Truly Matters

Priorities

In both our personal and professional lives, we face numerous decisions each day. Choosing one option often means excluding another—after all, you can't fully dedicate yourself to multiple large projects at once and maintain the same level of commitment. Although we may not consider it daily, these small decisions accumulate over time, shaping the trajectory of our careers. Do you remember the power of compounding discussed in Sect 7.4? So, how you can navigate this maze of daily tasks, many of which demand immediate attention, without losing sight of your most important long-term goals?

A key to success is recognising that effective time and career management requires more than just expertise in your field. One source of inspiration comes from Warren Buffet, the renowned investor and philanthropist. When asked by his pilot on how to achieve the success, Buffet advised him to create a list of the 25 most important goals in his life. From this list, he recommended selecting the top 5. He then explained that the remaining 20 goals, though they may seem important, actually serve as distractions from what truly matters.

What I'm trying to say here is that it is important to DECIDE what your priorities are. Do you know the true meaning of that word? It comes from two components in Latin: DE, meaning "off", and CAEDERE, meaning "to cut". "To decide" essentially means "to cut off". The original meaning of the word emphasises that a decision involves cutting off many possibilities in favour of one scenario. By deciding you eliminate everything that is not related to the choice you've made.

Eisenhower Matrix

But how do you determine which tasks are truly your priority? I turn to the matrix of General Dwight D. Eisenhower, the legendary commander and president of the United States. This simple method divides responsibilities into urgent and less urgent tasks, using a basic piece of paper divided into four quadrants.

- **Q1: Urgent and important tasks**—these are the things you must address immediately.
- **Q2: Important but less urgent tasks**—these are tasks you should plan to tackle soon.
- **Q3: Urgent but less important tasks**—these are typically tasks you can delegate or decline.

- **Q4: Non-urgent and unimportant tasks**—these are distractions that should be avoided.

To achieve maximum impact, focus on handling tasks in Q1 efficiently. Then, as soon as possible, shift your attention to Q2, where you should spend the bulk of your time. It's also essential to manage your activities and strategies so that Q3 actions are organized with minimal effort. Spending too much time in Q3 can sacrifice valuable time devoted to Q2 tasks. Regularly applying this matrix may feel challenging at first, but with practice, identifying important tasks versus less important ones will become second nature.

Hell Yes or No

For me, one of the most extreme yet effective ways to decide whether to take on a new project is the "Hell Yes or No" concept, as described by Derek Sivers in his book "Anything You Want". Sivers encourages to ask ourselves a single question when facing a new opportunity: Are you excited to dive in? If your immediate response is "maybe" or "yes, but," it's a clear sign that you shouldn't pursue it. This approach allows you to dedicate your time and energy to projects that genuinely excite and fulfil you. After all, isn't that what life is about? Pursuing what truly brings you joy and a sense of accomplishment. So, don't let small, insignificant tasks bog you down—focus on what truly matters to you.

Takeaway points:

- **You can't do everything—prioritisation is key.**
- **Prioritise your tasks based on urgency and importance.**
- **Say no to projects that don't feel right—protect your time for opportunities that truly align with your goals.**

GPSR Compliance

The European Union's (EU) General Product Safety Regulation (GPSR) is a set of rules that requires consumer products to be safe and our obligations to ensure this.

If you have any concerns about our products, you can contact us on

ProductSafety@springernature.com

In case Publisher is established outside the EU, the EU authorized representative is:

Springer Nature Customer Service Center GmbH
Europaplatz 3
69115 Heidelberg, Germany

www.ingramcontent.com/pod-product-compliance
Lightning Source LLC
Chambersburg PA
CBHW071605011225
36161CB00002B/41